Harvey Washington Wiley

Foods and food adulterants

Harvey Washington Wiley

Foods and food adulterants

ISBN/EAN: 9783337201098

Printed in Europe, USA, Canada, Australia, Japan

Cover: Foto ©Andreas Hilbeck / pixelio.de

More available books at **www.hansebooks.com**

U. S. DEPARTMENT OF AGRICULTURE.
DIVISION OF CHEMISTRY.
BULLETIN   No. 13.

FOODS

AND

FOOD ADULTERANTS.

INVESTIGATIONS MADE BY AUTHORITY OF
THE COMMISSIONER OF AGRICULTURE,
UNDER DIRECTION OF
THE CHEMIST.

PART FOURTH:
LARD AND LARD ADULTERATIONS,
BY
H. W. WILEY.

WASHINGTON:
GOVERNMENT PRINTING OFFICE.
1889.

[BULLETIN No. 13]

PART 4.—LARD AND LARD ADULTERATIONS.

U. S. DEPARTMENT OF AGRICULTURE,
DIVISION OF CHEMISTRY,
Washington, D. C., February 7, 1889.

SIR: With many interruptions, due to the experiments in the manufacture of sugar, carried on under the supervision of this division, I have completed our studies on lard and lard adulterations, and now have the honor to lay before you the results obtained for your inspection and approval.

I have endeavored to show the character of true lard, how it is made, and how it may be distinguished from its imitations. In the same manner the substances used in adulterating lard—viz, stearines and cotton oil—have been studied and their properties described. Also the characteristics of the mixed lards have been pointed out, and the best methods of analytical research illustrated.

Abstracts of similar studies by others have been given, and it is believed that the present state of our knowledge of lard and its compounds is fully set forth.

Some delay in submitting the manuscript to the Public Printer has been experienced on account of failure to arrange for printing the illustrations. To facilitate this matter, it has been decided to omit nearly all illustrations of methods of making and refining lard and cotton oil, and print only a few photo-micrographs showing the crystalline appearance of pure lard and stearines and mixtures thereof.

Respectfully,

H. W. WILEY,
Chemist.

Hon. NORMAN J. COLMAN,
Commissioner of Agriculture.

LARD AND LARD ADULTERATIONS.

(1) LARD.

(*a*) *Lard* is a term applied to the fat of the slaughtered hog, separated from the other tissues of the animal by the aid of heat.

In the crude state it is composed chiefly of the glycerides of the fatty acids, oleic and stearic or palmitic, with small portions of the connective tissues, animal gelatine, and other organic matters.

(*b*) *Kinds of lard.*—According to the parts of the fat used and the methods of rendering it lard is divided into several classes. According to methods of rendering lard is classified as kettle and steam. From material used the following classification may be made:

(*c*) *Neutral lard.*—Neutral lard is composed of the fats derived from the leaf of the slaughtered animal, taken in a perfectly fresh state. The leaf is either chilled in a cold atmosphere or treated with cold water to remove the animal heat. It is then reduced to a pulp in a grinder and passed at once to the rendering kettle. The fat is rendered at a temperature 105° to 120° F. (40°–50° C.). Only a part of the lard is separated at this temperature and the rest is sent to other rendering tanks to be made into another kind of product. The lard obtained as above is washed in a melted state with water containing a trace of sodium carbonate, sodium chloride, or a dilute acid. The lard thus formed is almost neutral, containing not to exceed .25 per cent. free acid; but it may contain a considerable quantity of water and some salt. This neutral lard is used almost exclusively for making butterine (oleomargarine).

(*d*) *Leaf lard.*—The residue unrendered in the above process is subjected to steam heat under pressure and the fat thus obtained is called leaf lard. Formerly this was the only kind of lard recognized in the Chicago Board of Trade, and was then made of the whole leaf.

(*e*) *Choice kettle-rendered lard; Choice lard.*—The quantity of lard required for butterine does not include all of the leaf produced. The remaining portions of the leaf, together with the fat cut from the backs, are rendered in steam-jacketed open kettles and produce a choice variety of lard known as "kettle-rendered." The hide is removed from the back fat before rendering and both leaf and back fat are passed through a pulping machine before they enter the kettle. Choice lard is thus defined by the regulations of the Chicago Board of Trade:

Choice lard.—Choice lard to be made from leaf and trimmings only, either steam or kettle rendered,—the manner of rendering to be branded on each tierce.

(*f*) *Prime steam lard.*—The prime steam lard of commerce is made as follows: The whole head of the hog, after the removal of the jowl, is used for rendering. The heads are placed in the bottom of the rendering tank. The fat is pulled off of the small intestines and also placed in the tank. Any fat that may be attached to the heart of the animal is also used. In houses where kettle-rendered lard is not made the back fat and trimmings are also used. When there is no demand for leaf lard the leaf is also put into the rendering tank with the other portions of the body mentioned. It is thus seen that prime steam lard may be taken to represent the fat of the whole animal, or only portions thereof. The quantity of fat afforded by each animal varies with the market to which the meat is to be sent. A hog trimmed for the domestic market will give an average of about 40 pounds, while from one destined for the English market only about 20 pounds of lard will be made. Prime steam lard is thus defined by the Chicago Board of Trade:

Prime steam lard.—Standard prime steam lard shall be solely the product of the trimmings and other fat parts of hogs, rendered in tanks by the direct application of steam, and without subsequent change in grain or character by the use of agitators or other machinery, except as such change may unavoidably come from transportation. It shall have proper color, flavor, and soundness for keeping, and no material which has been salted shall be included. The name and location of the renderer and the grade of the lard shall be plainly branded on each package at the time of packing.

This lard is passed solely on inspection; the inspector having no authority to supervise rendering establishments in order to secure a proper control of the kettles. According to the printed regulations, any part of the hog containing fat can be legally used.

Since much uncertainty exists in regard to the disposition which is made of the guts of the hog I have had the subject carefully investigated. Following are the results of the study:

(*g*) *Guts.*—The definition of the term as used by hog packers is: Everything inside of a hog except the lungs and hearts, or, in other words, the abdominal viscera complete. The material is handled as follows:

When the hog is split open the viscera are separated by cutting out the portion of flesh surrounding the anus and taking a strip containing the external urino-generative organs. The whole viscera are thrown on a table and divided as follows: The heart is thrown to one side and the fatty portion trimmed off for lard. The rest goes into the offal tank or sausage. The lungs and liver go into the offal tank (or sausage).

The rectum and large intestines are pulled from the intestinal fat and peritoneum and, along with the adhering flesh and genito-urinary organs, sent to the trimmer. All flesh and the above-mentioned organs are trimmed off and the intestine proper is used for sausage casings. The trimmings, including the genito urinary organs, are washed and dumped into the rendering tank. The small intestine is also pulled from the fatty membrane surrounding it and saved for sausage casings. The remaining material, consisting of the peritoneum, diaphragm, stomach, and adhering membranes, together with the intestinal fat, constitute the "guts" which are seen undergoing the process of washing, which is usually conducted in three or four different tanks. As the "guts" pass into the first tank the stomach and peritoneum are split open and also any portion of the intestines which sometimes adhere to the peritoneum. After receiving a rough wash they are passed from tank to tank, when, after the third or fourth wash, they are ready for the rendering tank. The omentum fat is cut from the kidneys and the kidneys with a little adhering fat go into the rendering tank. Spleen and pancreas go into the rendering tanks, as do also the trachea, vocal chords, and œsophagus.

To sum up, it is safe to say that everything goes into the rendering tank, with the following exceptions:

(1) The intestines proper, which are saved for sausage-casings.
(2) The liver and lungs.
(3) That part of the heart free from fat.

I have been told that in killing small hogs, and also when there is small demand for sausage-casings, it is frequently the practice to split the intestines, so as to save expense of pulling from the fat, and after washing, fat and all go into the tank. Of course it will often happen that the intestines break off and portions adhere to the enveloping tissue, and consequently get into the tank after washing.

It is a commercial fact that sausage-casings are worth more than the small amount of adhering fat, and consequently packers will save them. Small hogs produce small casings difficult to pull, and it is reasonable to believe that they will be handled in the simpler manner. They break so easily that they are hardly worth saving separately. It is stated by lard manufacturers that the grease made from the parts of the intestines mentioned above is used for the manufacture of lard oil and soap, and does not enter into the lard of commerce.

(*h*) *Butchers' lard.*—The small quantities of lard made by butchers are usually "kettle-rendered," after the manner practiced by small farmers in making lard for home consumption. Often the scraps are saved up for a considerable length of time by the butchers before rendering, and that is likely to increase the free acid present. This lard is also frequently dark colored, and contains a considerable quantity of glue. In New York this lard is known as "New York City Lard."

In this figure is represented the type of apparatus used for rendering lard, etc., under pressure. The rendering vessel is made of boiler iron or steel, and varies in size according to the magnitude of the establish-

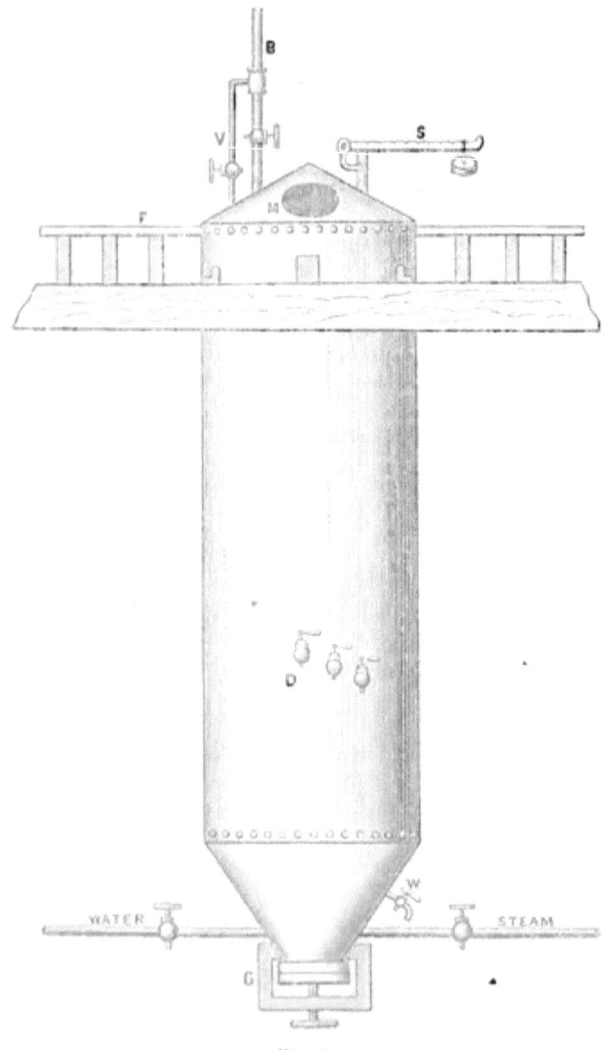

Fig. 16.

ment. A very common size is 10 to 12 feet in length and 3 to 5 feet in diameter. The heads, scraps, and other materials are put in at M. When the tank is full M is closed. Steam is admitted through the pipe thus marked, and condensed water drawn off through the water-pipe. Through the cocks at D the depth of lard in the tank can be determined and the lard drawn off. When the process is finished and the lard drawn off the bottom G is opened and the "tankage" withdrawn and dried for fertilizing purposes.

(B) Other Hog-Fat Products.

There are many other hog-fat products not used in the manufacture of lard or compound lard, a description of which, however, may prove useful here.

(a) *White grease.*—This grease is made chiefly from hogs which die in transit, by being smothered or frozen. Formerly it was also made from animals dead of disease; but this product has of late been diminished on account of certain State laws requiring the carcasses of hogs which have died of cholera to be buried. This grease is made from the whole animal with the exception of the intestines. The latter are rendered separately and make "brown grease". The rendering is done in closed tanks at a high pressure. The residue is used in the manufacture of fertilizer. White and brown grease are used chiefly in the manufacture of low-grade lard oils and soap.

(b) *Yellow grease.*—Yellow grease is made by packers. All the refuse materials of the packing-houses go into the yellow-grease tank, together with any hogs which may die on the packers' hands. Yellow grease is intermediate in value between white and brown. It is used for the same purposes.

(c) *Pigs'-foot grease.*—This grease is obtained chiefly from the glue factories, and is used for making lard oils and soap.

(2) STEARINES.

The stearines are the more solid portions of the animal fats remaining after the more fluid portions have been removed by pressure. The stearines used in the manufacture of compound lard are lard stearine, derived from lard, and oleo stearine, derived from a certain quality of beef tallow. Cotton-oil stearine is used chiefly in the manufacture of butterine.

A.—Lard Stearine.

The lard stearine used in compound lard is made as follows:

The prime steam lard, if properly crystallized and of the right temperature (from $45°$ to $55°$ F., winter; $55°$ to $65°$, summer), is sent at once to the presses. If not properly grained, it is melted and kept in a crystallizing room at $50°$ to $60°$ F., until the proper grain is formed. The lard is then wrapped in cakes with cloth, each cake containing 10 to 20 pounds. The cakes are then placed in a large press, with suitable septa to facilitate the egress of the oil. These presses are sometimes 40 to 50 feet in length, and when first filled 12 to 18 feet high. The pressure is applied very gradually at first by means of a lever working a capstan, about which the chain is wrapped, attached to the upper movable part of the press.

The oil expressed, prime or extra lard oil, is used for illuminating and lubricating purposes. The resulting stearine is used for making compound lard and is worth more than the lard. It has about .5 per cent. free fatty acid (less than the lard oil), and crystallizes in long needles, making the texture tough.

B.—Oleo-Stearine.

This product is made chiefly from the caul fat of beeves. This fat is rendered in open kettles at a low temperature. The resulting tallow is placed in cars in a granulating room, where it is allowed to remain for thirty-six to forty-eight hours at a temperature 80° to 90° Fah. The contents of the cars are then mixed and placed on a revolving table, where they are made into cakes. These are wrapped with strong cotton cloth and placed in a strong press, where a gradual pressure at 90° F., becoming very strong at the end, is applied for one or two hours. The expressed oil, known as oleo-oil, is used in the manufacture of butterine. The stearine is removed from the press as white hard cakes, and is used for adulterating lard. The oil is sometimes filtered with a small percentage of fuller's earth, to improve its color and brightness.

C.—Mutton Tallow.

A fine article of mutton tallow is also sometimes used in lard, but the objection to the flavor is sufficient to limit its use to a small amount.

D.—Beef Fat.

The following general remarks on beef fat will be found instructive:

Before the day of the oleomargarine industry all fat rendered from the tissues of cattle was known commercially as tallow. Since then differentiation has taken place and the term tallow is no longer sufficient to designate the several products obtained from the rendered fat of the beef. We have first "butter stock," which is rendered from the caul fat at a low temperature and from which is manufactured by means of pressure—

(1) Oleo-oil.

(2) Oleo-stearine (beef stearine).

The kidney fat as a rule is left with the carcass and constitutes what is known as suet. Marrow stock, as its name implies, is rendered marrow fat, and when properly prepared is almost equal to butter stock in quality. Tallow is made from the trimmings and portions of the viscera. Its color varies from white to yellow according to the portions of the animal which have been used and the care with which they have been prepared for rendering and the temperature at which rendered. When freshly and carefully rendered tallow should show less than 1.5 per cent. of free fatty acid. The tallow on the market will show anywhere from 2 to 10 per cent. Its flavor varies, never being good enough for lard. Tallow grease corresponds to the yellow grease of the hog-packer. It is of a dark color and often contains as much as 50 per cent. of free acid. It is made into low-grade soaps.

(3) COTTON OIL.

(a) The cotton seed from various sources is put through a screen to take out the bolls and coarse material. The seed is then put through a gin to remove as far as possible any remaining lint, of which about 20 pounds per ton of seed are obtained.

The clean seed is next sent to a huller composed of revolving cylinders covered with knives, which cut up both seed and hull. The chips are then conveyed to a screen placed on a vibrating frame, through which the kernels fall. The hulls are carried by an endless belt to the furnaces, where they are burned. The kernels of the seed are conveyed to crusher rolls, where they are ground to a fine meal. The meal is then sent to a heater, where it remains from twenty to forty minutes. These heaters have a temperature of 210° to 215° F. The hot meal is formed into cakes by machinery; these are wrapped in cloth and placed in the press. About 16 pounds of meal are put in each cake. The cakes are placed in a hydraulic press, where a pressure of from 3,000 to 4,000 pounds per square inch is applied. The press is also kept warm. The expressed cakes contain only about 10 per cent. of oil. The cake is sold as cattle food or for fertilizing purposes. The crude oil as thus expressed contains about 1.5 per cent. of free acid. The chief cotton-seed presses of the country are located at the following points:

Cotton-seed oil milling points.

Arkansas:
  Little Rock.
  Argenta.
  Fort Smith.
  Texarkana.
  Brinkley.
  Helena.
Alabama:
  Selma.
  Mobile.
  Montgomery.
  Eufaula.
  Huntsville.
Georgia:
  Atlanta.
  Augusta.
  Albany.
  Columbus.
  Macon.
  Rome.

Illinois:
  Cairo.
Louisiana:
  New Orleans.
  Shreveport.
  Baton Rouge.
  Monroe.
Missouri:
  Saint Louis.
Mississippi:
  Clarksdale.
  Columbus.
  Canton.
  Grenada.
  Greenville.
  Meridian.
  Natchez.
  Vicksburg.
  West Point.

North Carolina:
  Charlotte.
  Raleigh.
Tennessee:
  Memphis.
  Jackson.
  Nashville.
  Dyersburgh.
Texas:
  Brenham.
  Dallas.
  Galveston.
  Houston.
  Palestine.
  Waco.

The oil is chiefly pressed in winter, since it is difficult to keep the seed for summer work. Some mills are, however, operated during the summer. The crude oil is shipped in tanks holding from 36,000 to 45,000 pounds each. When the oil is shipped North in winter it usually becomes solidified. In order to get it out of the tanks they are placed on switches and a jet of steam is introduced into the tank and the oil gradually melted out. Another method consists in covering the tank with wood, forming a chamber into which exhausted steam is introduced. Gutters are provided along the railroad tracks into which the oil flows and is conducted into the receiving tanks. From the receiving tanks it is pumped into large receivers called scale tanks, where the crude oil is weighed.

(b) *Refining process.*—After weighing, the oil is pumped into refining kettles. These are of various sizes, the largest ones being 20 to 25 feet deep and 15 feet in diameter. These tanks are furnished with steam-coils for the purpose of heating the oil and with appropriate machinery for keeping it in motion. A solution of caustic soda is used for refining. This solution is made from 10° to 28° Beaumé in strength, and varying quantities are used according to the nature of the oil operated upon. After the addition of the caustic soda the mixture is agitated for forty-five minutes and kept at a temperature of 100° to 110° F. The contents of the tank are then allowed to stand six to thirty-six hours, when the solid matters, soap and substances precipitated by the caustic alkali gather at the bottom. This mixture is called "foots," and is used for making soap. The yellow oil resulting by this process is further purified by being heated and allowed to settle again or by filtration and is called summer yellow oil. Winter yellow oil is made from the above material by chilling it until it partially crystallizes and separating the stearine formed, about 25 per cent., in presses similar to those used for lard. This cotton-oil stearine is used for making butterine and soap.

(c) *White oil.*—The yellow oil obtained as above is treated with from 2 to 3 per cent. of fuller's earth in a tank furnished with apparatus for keeping the mixture in motion. When the fuller's earth has been thus thoroughly mixed with the oil, the whole is sent to the filter press. The fuller's earth has the property of absorbing or holding back the yellow coloring matter, so that the oil which issues from the press is almost white. This white oil is the one which is chiefly used for making compound lard.

Cotton oil is obtained from the seeds of *Gossypium herbaceum*. The percentage of oil varies in the seed from 10 to 30.

In 1882 it was estimated that the oil industry was represented by the following data:*

110,000 tons of seed, yielding 35 gallons of crude oil to the ton, are 11,550,000 gallons, worth 30 cents per gallon $1,365,000
Same amount of seed, yielding 22 pounds cotton lint to the ton, is 2,020,000 pounds cotton, worth 8 cents per pound 724,000
And yielding also 750 pounds of oil-cake to the ton (2,240 pounds) is 1,37,277 tons of cake at $20 per ton 2,745,540
                                                                        7,772,140
Deduct the sum paid for the seed, say 4,104,000

From September 1, 1883, to September 1, 1886, there were exported from New York 88,871 barrels, and from New Orleans 186,720 barrels, making a total of 275,591 barrels from the two ports. These figures show conclusively that American cotton-seed oil is growing rapidly in favor in foreign countries.

When well stored and properly ventilated, cotton seed keeps sweet for twelve months. If allowed to become damp, or stored too long in bulk, it grows heated, and is liable to spontaneous combustion.

Manufacture of cotton-seed oil.—The seed when landed at the mill is first examined. If too damp or wet it is dried by spreading it over a floor with free access of air, exposing it on frames to the sunlight in warm weather, or by kiln-drying. Drying is the exception rather than the rule in the United States. Cotton ginning is so carefully done that the seeds have little or no opportunity to become wet. Besides this, the seed is generally held at the gins for some time before it is sold to the oil manufacturer.

The first process in preparing the dry seed for the mill is to free it from dust. This is effected by shaking it in a screen or in drums lined with a fine metallic net and containing a strong magnet to which any iron nails will adhere, which are frequently present. From the drums the seeds drop into a gutter leading to a machine which removes the lint left by the gin. This is done by a gin constructed for the purpose, with saws closer together than the ordinary cotton-gin. An average of twenty-two pounds of short lint is taken from a ton of the seed. This product, called "linters," is used in the manufacture of cotton batting. The clean seeds are then transferred to the sheller, which consists of a revolving cylinder containing twenty-four cylindrical knives and four back knives. The sheller revolves at great speed, and as the seed is forced between the knives the pericarp or hull is broken and forced from the kernel. The mixed shells and kernels are separated in a winnowing machine by a strong blast of air. This removal of the husk makes a vast difference in the meal cake, a dessicated or decorticated cake being five times more nutritious and wholesome than an undecorticated cake.

Being thus cleaned, shelled, and separated, the kernels are carried by a system of elevators to the upper story and then pass down into the crusher-rolls to be ground to flour.

Cold pressure produces a very good salad oil, and this is the method generally pursued in Marseilles and other European cities for the first pressure, after which the residue is subjected to a second warm pressure. In this country, however, warm pressure is generally preferred. The meal is heated in a meal heater for fifteen to twenty minutes to 204.4° to 215.3° F.

The heated meal is placed in woolen bags, each holding sufficient seed for a cake. The bags are then placed between horse-hair mats backed with leather having a fluted surface inside to facilitate the escape of the oil under the hydraulic pressure amounting to 169 tons. With the most improved presses the hair mats are, however, done away with. The bags remain in the press seventeen minutes, the solid "oil-cake" of commerce remaining behind. This cake forms a superior feed for cattle, horses, sheep, and especially swine, and is nutritious, easily digested, and fattening.

Cotton-seed cake is of a rich golden color, quite dry, and has a sweet, nutty, oleaginous taste. When ground to the fineness of corn meal it is known as "cotton-seed meal," and in that form is frequently used for fertilizing purposes.

The crude oil as obtained from the press is pumped into the oil-room and either barreled for shipment or refined.

Four qualities of the oil are known:

Crude oil is thickly fluid and of a dirty yellow to reddish color; on standing it deposits a slimy sediment. The *second* quality has a pale orange color and is obtained by refining the crude oil. The *third* quality is obtained by further purification of the second; and the *fourth*, which has a pale straw color and a pure nutty taste, by bleaching the third quality.

The coloring principle, termed *gossypin*, is collected on a filter, carefully washed to remove any trace of acid, and dried slowly at a low temperature. It is then ready for use as a dye, and gives fast colors on both silk and wool. It is claimed that the quantity of coloring matter in a ton of crude oil is 15 pounds, though this proportion must vary considerably. Its properties are insolubility in acids, slight solubility in water, free solubility in alcohol or alkalies. In its dry state it is a light powder of a pungent odor, of a brown color, and strongly tinctorial.

Crude cotton-seed is thickly fluid, twenty-eight to thirty times less fluid than water, and has a specific gravity of 0.9283 at 68° F., 0.9306 at 59° F., and 0.9343 at 50° F.

According to the quality of the oil, palmitin is separated between 54° and 43° F. The oil congeals at 28.5° to 27° F. In taste and odor it resembles linseed oil, and as regards other properties it is an intermediate between drying and non-drying oils.

Refined cotton-seed oil has a specific gravity of 0.9264 at 59° F.; it separates palmitin already below 53.5° F., and congeals at 32° to 30° F.

The oil consists of palmitin and olein, and to make it still more adapted for the adulteration of olive oil, for which immense quantities are used every year, it is intentionally cooled for the separation of palmitin, which lowers the specific gravity.

MIXING.

The term refined lard has long been used to designate a lard composed chiefly of cotton oil and stearine. The largest manufacturers of this kind of lard have now abandoned this term and are using the label "lard compound" instead. This is but just to the consumers of this article who are likely to be misled by the term refined lard. The prime steam lard in a state of fusion, the stearine also in a liquid condition, and the refined cotton oil are measured in the proportions to be used and placed in a tank at a temperature of 120° to 160° F. In this tank the ingredients are thoroughly mixed by means of paddles operated by machinery. After mixing the compound lard passes at once to artificial coolers where it is chilled as soon as possible. It is thence run directly into small tin cans or large packages and prepared for market.

(1) PROPERTIES OF PURE LARD.

A.—PHYSICAL PROPERTIES.

(*a*) *Specific gravity.*—The specific gravity of a pure lard varies rapidly with the temperature. It is not convenient to take the specific gravity of a lard at a lower temperature than 35° or 40°,* inasmuch as below that temperature solidification is apt to begin. The specific gravity, therefore, is usually taken at 35° or 40° or at the temperature of boiling water, viz., 100°. At 40° the specific gravity of pure lard is about .890, and at 100° about .860, referred to water at 4°.

The specific gravity of pure lard does not differ greatly from that of many of the substances used in adulterating it, but it is distinctly lower than that of cotton oil, and is of great distinctive value in analysis.

(*b*) *Melting point.*—The melting point of a pure lard is a physical

---

* All degrees are Centigrade unless otherwise stated.

characteristic of great value. The melting point of the fat of the swine varies with the part of the body from which it is taken. The fat from the foot of the swine appears to have the least melting point, viz, 35.1°. The intestinal fat seems to have the highest, viz, 41°. In fat derived from the head of the animal the melting point is found to be 35.5°, while the kidney fat of the same animal shows a melting point of 42.5°. In steam lards, representing the lards passed by the Chicago Board of Trade, the melting point for ten samples was found to vary between 29.8° and 43.9°. In general it may be said that the melting point of steam lards is about 37° which is the mean of ten samples examined. In pure lards derived from other localities the melting point was also found to vary. A sample of lard from Deerfoot Farm, Southborough, Mass., was found to have a melting point of 44.9°, while a pure lard from Sperry & Barnes, New Haven, Conn., melted at 39°. The mean for eighteen samples was 40.7°. While the melting point can not be taken as a certain indication of the purity of a lard, nevertheless a wide variation from 40° in the melting point of a lard should lead at least to a suspicion of its genuineness, or that it was made from some special part of the animal. Perhaps one reason why the melting point has not been more highly regarded by analysts is because of the unsatisfactory method of determining it; but when it is ascertained by the method used in these investigations it becomes a characteristic of great value.

(c) *Color reaction.*—The coloration produced on pure lard by certain reagents serves as a valuable diagnostic sign in the analysis of lard and its adulterations. Various reagents have been employed for the production of characteristic colors in fats, but of these only two are of essential importance. They are sulphuric and nitric acids. Pure lard, when mixed with sulphuric and nitric acids of the proper density, as indicated hereafter, give only a slight color which varies from light pink to faint brown. The variation produced in the colors by pure lards is doubtless due to the presence in various quantities of certain tissues of the animal other than fat. For instance, a variation in the amount of gelatinous substance mechanically entangled with the lard or of the tissues composing the cells in which the lard was originally contained would be entirely sufficient to account for the slight differences in color produced by lards of known purity. It might, therefore, be difficult to distinguish accurately between a pure lard containing a considerable amount of other tissues from the animal and one which contained a small amount of adulteration. The coloration produced, therefore, by the acids named should not be relied upon wholly in distinguishing pure and adulterated lards; but the character of such coloration should be carefully noted in the analyst's book. In the steam lards examined some of the remarks describing the coloration produced are as follows:

"'Trace of color,' 'faint pink,' 'bright pink,' 'light red,' 'yellowish,' etc. For pure lards of miscellaneous origin some of the descriptions are as follows: "Brownish pink," "trace of yellow," "marked red brown," "no color," "slight coloration, etc."

(d) *Refractive index.*—The deviation produced in the direction of a ray of light in passing through a film of melted fat is also a valuable physical characteristic. This deviation is usually measured as the quotient of the sine of the angle of incidence divided by the sine of the angle of refraction and is known as the refractive index. The refractive index of pure water, at 25° on the instrument used in these investigations was 1.3300. The refractive index of the samples of lard was made at as low a temperature as possible to preserve fluidity, viz: between 30° and 36°. In the tables the temperature at which the index was taken is not given, but the number representing the index corrected to the uniform temperature 25°. The rate of variation in the refrative index for each degree of temperature, experimentally determined, for lard oil was .000288. This number may also be taken to represent the variation for lard. The refractive index varies inversely as the temperature. The mean number for a pure lard at 25° is about 1.4620. The variation from this number can be seen in the analytical tables which follow. The refractive index of pure lard is distinctly less than that of cotton-seed oil at the same temperature, and is therefore a valuable characteristic for analytical purposes.

(e) *Rise of temperature with sulphuric acid.*—More valuable for diagnostic purposes than the physical properties already considered is the rise of temperature which lard undergoes when mixed, under proper conditions, with sulphuric acid. There is such a marked difference between the numbers representing the rise of temperature in pure lard and those of the adulterants usually employed in the manufacture of mixed lard as to give this number a high analytical value. With steam lards, ten samples, the extremes, as registered by the thermometer, were 38.8° and 42.1°. For pure lards of miscellaneous origin, one from Deerfoot Farm, Southborough, Mass., gave a rise of temperature 37.1°, and a pure leaf lard from Sperry & Barnes, New Haven, Conn., a rise of temperature of 46.2°.

The value of this characteristic is so great as to lead me to expect approximately reliable quantitative results from a general determination of the actual amount of heat produced in an appropriate calorimeter. I am at present attempting to devise an instrument by which the actual number of calories produced by mixing definite quantities of fats and oils and sulphuric acid can be accurately determined.

(f) *Crystallization point of fatty acids.*—The method described in the work of Dalican for determining the crystallizing points of fatty acids gives valuable data concerning the nature of pure lard, and also of the relative amount of stearic and oleic acids present in the mixture. The crystallizing point was found to vary in the ten samples of prime steam lard already mentioned from 35.4° to 39.5°. In pure lards of other kinds the variation was found to be from 32.1° to 42.7°.

(*g*) *Melting point of fatty acids.*—In connection with the crystallizing point of the fatty acids, the melting point is also of value. This temperature has been determined in the fat acids derived from steam and pure lards, and the numbers will be found in the analytical tables. In the prime steam lards these numbers vary from 41.4° to 43°. In pure lards of other kinds the variation was from 36.9° to 46.6°.

B.—CHEMICAL PROPERTIES.

(*a*) *Volatile acids.*—The quantity of volatile acid, as ordinarily estimated in a pure lard, is quite minute. Unless some suspicion of adulteration is awakened the search for such volatile or soluble acid may be omitted. Measured by the decinormal alkali solution required for 5 grammes of the fat the mean quantity of volatile acid in a pure lard may vary from .2 to .4 of a cubic centimeter. The determination, therefore, of the volatile acid in the examination of lards has none of that high diagnostic value which attaches to it in the examination of butters.

(*b*) *Fixed acids.*—The quantity of fixed acids (non-volatile and insoluble in water) in lard varies from 93 to 95 per cent.

(*c*) *Free acids.*—The quantity of free acids in lard rarely exceeds .5 per cent.

Twelve determinations of free acids in lards of known purity gave the following numbers expressed as per cent.:

.54   .92   .55   .75   .75   .35   .65   .60   .45   1.0   .40   .50

(*d*) *Saponification equivalent.*—The amount of caustic alkali necessary to saponify the fatty acids of the common glycerides is known as its saponification equivalent or number. The operation is usually known as Koettstoffer's process. The number of parts of a glyceride saponified by one equivalent of alkali is represented by one-third of the molecular weight of the glyceride in question. The saponification equivalent, therefore, represents the number of grams of an oil or fat saponified by one equivalent in grams of an alkali. The percentage of caustic potash used for saponifying a lard is about 20 and the mean saponification equivalent about 285. In the prime steam lards examined by us, the extreme variations were 276.14 and 290.05, and the mean 283.45. In pure lards of other kinds the extremes were 272.64 and 294.14, and the mean 280.33.

(*e*) *Iodine number.*—The quantity of iodine absorbed by an oil or fat affords one of the most valuable indications of its constitution. The glycerides of the olein series have the property of absorbing the halogens. On the other hand the glycerides of the stearic series do not absorb iodine. Hence in a fat or oil from which the presence of linolein and its analogous bodies can be excluded the quantity of iodine absorbed may become a fairly accurate measure of the amount of oleic acid present. The lard derived from different portions of the swine varies largely in the amount of olein contained therein. For instance, a sample of intestinal lard absorbed 57.34 per cent. of iodine; the leaf lard from the same animal absorbed 52.55 per cent., the foot lard 77.28 per cent., the head lard 85.03 per cent. In the prime steam lards mentioned the variation in the percentage of iodine absorbed was from 60.34 to 66.47 per cent., and the mean 62.86 per cent. In pure lards of other kinds the mean was 62.48 per cent. Thus in lards of known purity the amount of iodine absorbed will indicate the probable part of the animal from which the fat in the lard was derived. The wide variation between the iodine numbers of pure lard and those of the adulterants used in making compound lard serve to render this number of the greatest importance in analytical work.

(*f*) *The reaction with nitrate of silver.*—Pure lards, treated with a solution of nitrate of silver, after the method of Bechi, or the fatty acids thereof, after the method of Milliau, give no reduction of metallic silver, or, at most, only a trace and no or only a slight coloration. This fact is of the utmost importance in the analysis of lard.

(*g*) *Microscopical appearances.*—Lard, examined with the microscope, shows a definite crystalline structure, but does not plainly reveal the character of the crystals. When lard is slowly crystallized from ether, beautiful rhombic crystals of stearine are obtained, which are easily distinguished from the groups of fan-shaped crystals given by beef or mutton fat under similar conditions.

(*h*) *Moisture in lard.*—The quantity of water in pure lard varies from a mere trace to .7 per cent. Twelve determinations showed the following per cents.:

.7 .4 .2 .5 .6 .5
.2 .2 .3 .3 .3 .7

(5) PROPERTIES OF LARD ADULTERANTS.

COTTONSEED OIL.

A.—PHYSICAL PROPERTIES.

(*a*) *Specific gravity.*—Cottonseed oil being liquid at ordinary temperatures, its specific gravity can be easily taken at the temperature of the room. For purposes of comparison, the rate of variation in the specific gravity of the oil can be determined and its specific gravity at any given temperature calculated, or its specific gravity can be directly determined at $35°$, $40°$, or $100°$, as may be desired, by comparison with water at the same temperature. In the samples examined the specific gravities of the oils at $35°$ vary from .9132 to .9154. The mean for nineteen samples is .9142. These numbers show the relative weight of the oil, an equal volume of water at the same temperature being taken as unity.

LARD AND LARD ADULTERATIONS.

Specific gravity of refined cotton oil at different temperatures.

[Water at 15° C.=1. Average, oil at 15°=.9218—at 100° .8685.]

Tempera-ture C°.	Specific gravity.	Weight cubic foot oil.	Tempera-ture C°.	Specific gravity.	Weight cubic foot oil.
		Pounds.			*Pounds.*
10	.9249	57.68	56	.8958	55.86
11	.9243	57.63	57	.8953	55.82
12	.9237	57.60	58	.8946	55.78
13	.9231	57.55	59	.8940	55.74
14	.9224	57.51	60	.8934	55.71
15	.9218	57.49	61	.8927	55.67
16	.9212	57.45	62	.8921	55.63
17	.9206	57.41	63	.8915	55.59
18	.9199	57.35	64	.8908	55.55
19	.9193	57.31	65	.8902	55.51
20	.9187	57.28	66	.8896	55.54
21	.9181	57.25	67	.8890	55.48
22	.9174	57.21	68	.8883	55.39
23	.9168	57.18	69	.8877	55.35
24	.9161	57.13	70	.8870	55.31
25	.9155	57.08	71	.8864	55.28
26	.9149	57.04	72	.8858	55.25
27	.9143	56.00	73	.8851	55.20
28	.9136	56.97	74	.8845	55.16
29	.9129	56.93	75	.8839	55.13
30	.9123	56.89	76	.8832	55.19
31	.9117	56.85	77	.8826	55.05
32	.9110	56.81	78	.8820	55.01
33	.9105	56.75	79	.8814	54.98
34	.9098	56.74	80	.8807	54.94
35	.9092	56.70	81	.8801	54.90
36	.9086	56.66	82	.8795	54.86
37	.9079	56.61	83	.8788	54.80
38	.9073	56.58	84	.8782	54.76
39	.9067	56.54	85	.8776	54.73
40	.9060	56.50	86	.8769	54.68
41	.9054	56.46	87	.8763	54.64
42	.9048	56.42	88	.8757	54.60
43	.9043	56.38	89	.8751	54.56
44	.9035	56.34	90	.8744	54.53
45	.9029	56.30	91	.8738	54.49
46	.9022	56.26	92	.8732	54.45
47	.9016	56.23	93	.8725	54.40
48	.9010	56.18	94	.8719	54.36
49	.9004	56.13	95	.8712	54.31
50	.8997	56.10	96	.8706	54.28
51	.8991	56.07	97	.8700	54.24
52	.8984	56.03	98	.8695	54.20
53	.8978	55.99	99	.8689	54.16
54	.8972	55.95	100	.8683	54.10
55	.8966	55.91	....	....	....

(b) *Melting point.*—Since cotton oil solidifies only at a temperature near or below the freezing point of water its melting point has not been determined.

(c) *Color reaction.*—The color produced in cotton oil by sulphuric and nitric acids is a characteristic mark of the greatest value. This color varies from deep reddish brown to an almost black color. Some of the descriptions of the color produced in cotton oil, taken from the note-book, are as follows: "dark brown," "very brown black," "deep red brown," "very red," "yellow brown," etc. It must not be forgotten, however, that these colors can be produced by other oils, and hence their occurrence is not conclusive evidence of the presence of cotton oil.

(d) *Refractive index.*—The refractive index of cotton oil is distinctly higher than that of lard. The variation in the index of refraction is inversely as the temperature. The mean rate of variation for each degree is .000288. For a temperature of 25° the mean refractive index of the samples examined was 1.4674. The rate of variation in the index of refraction in cotton oil is sensibly the same as that for lard.

(e) *Rise of temperature with sulphuric acid.*—The rise of temperature which cotton oil suffers when mixed with sulphuric acid is a very prominent diagnostic sign. In the samples examined the lowest increment of temperature noted was 80.4° and the highest 90.2°. The mean rise of temperature was 85.4°. Cotton oil, therefore, gives more than double the increment of temperature shown by pure lard under the same conditions.

(f) *Crystallization point of fatty acids.*—Since cotton oil is fluid even at low temperatures (viz, 0°) the determination of its melting point is only a matter of scientific interest. The point at which its free acids crystallize is, however, easily determined according to the method of Dalican.

	Degrees
The mean crystallizing point of the acids examined was	33.5
The minimum was	30.5
The maximum was	35.6

The high temperature reached in the crystallization of the fat acids is a peculiar characteristic of cotton oil. In lard there is not a very great difference between the temperatures indicated by the melting point of the glycerides and the crystallizing point of the fat acids. In cotton oil, however, these temperatures are widely removed.

(g) *Melting point of fatty acids.*—The melting point of the free acids of cotton oil was determined both in capillary tubes and by observing the deportment of the acid on the bulb of a delicate thermometer protected by a glass flask. The two sets of data were almost identical.

	Degrees
The mean melting point of the acids examined was	39.4
Maximum	44.4
Minimum	34.6

The characteristics mentioned above are emphasized when the melting point of the fat acids is considered. These numbers seem much higher than would be expected.

B.—CHEMICAL PROPERTIES.

(a) *Volatile acids.*—The statements made in regard to the volatile acids in a pure lard are also applicable to cotton-seed oil.

For 5 grammes of cotton oil the quantity of deci-normal alkali consumed is slightly greater than for pure lard and may amount to as much as .5 cc.

If cocoa oil is present the number will be much higher. 5 grammes of pure cocoa oil will consume from .7 to .8 cc of the deci-normal alkali.

(b) *Saponification equivalent.*—In the samples reported the mean saponification equivalent was 283.8, although in some instances quite a difference was noticed from this figure.

(c) *Iodine number.*—Cotton oil possesses in a much higher degree than lard the property of absorbing iodine. This is due not only to the large percentage of oleic acid which it contains, but also probably to the presence of a small amount of linoleic acid or some homologue thereof. In the samples examined in no case did the iodine number fall below 100 and in one instance it rose to 116.97. The mean iodine number was 109.02.

(d) *Reaction with nitrate of silver.*—A more important property even than its power of absorbing iodine is shown by cotton oil in the reduction of silver to the metallic state under certain conditions. The test may be applied, as already indicated, either to the oil itself or to the fatty acids thereof. The silver is either reduced in the form of a metallic mirror deposited on the sides of the vessel or in minute black particles which give a brown or black appearance to the liquid. In some cases the liquid shows a greenish tint.

OTHER PROPERTIES.

The refined cotton oil used in adulterating lard has a pleasant taste, is almost odorless, and possesses a faint yellow color. Its resemblance to olive oil is so marked that for all culinary purposes it forms an excellent substitute therefor. Cotton oil possesses slight drying qualities which render it unfit for lubricating delicate machinery. Therefore it can never take the place of sweet oil for that purpose.

STEARINES.

The stearines used in the adulteration of lard are derived chiefly from lard, certain parts of beef fat, and cotton oil. These are generally called lard stearine, oleo-stearine, and cotton-oil stearine, respectively.

A.—PHYSICAL PROPERTIES.

(*a*) *Specific gravity.*—The specific gravity of stearines may be taken in their solid state or in a liquid state at a high temperature, 40° to 100°.

(*b*) *Melting point.*—The melting points of the stearines are higher than the natural glycerides from which they are derived. A prime oleo-stearine from Armour & Co., Chicago, showed a melting point of 51.9°. A prime lard stearine from the same firm showed a melting point of 44.3°, which is only slightly higher than the mean melting point of pure lards. The lowest melting point of any stearine examined was a sample of dead-hog stearine from J. P. Squire, Boston, which was 38.2°. The highest observed melting point in the stearines examined was an oleo-stearine from N. K. Fairbank & Co., Chicago, showing 53.8°. The high melting point of the stearines is a characteristic of great value in the adulteration of lard since it serves to counteract the influence of the cotton oil, which of course tends to lower the melting point of any lard mixture into which it may enter. The influence of the various constituents, however, on the melting point does not seem to be proportional to the respective quantity of each therein. For instance, a mixture of 25 per cent. of cotton oil having a melting point below zero, with 25 per cent. of an oleo-stearine having a melting point of only about 12° above the normal for pure lard, with 50 per cent. of pure lard of normal melting point, might not show a lowering of the melting point at all proportional to the presumable influence of the cotton oil present. The cotton-oil stearine, as might be expected, has a melting point below that of the similar products derived from lard and tallow.

(*c*) *Color reaction.*—The color reactions produced in the stearines by sulphuric and nitric acids are much the same as those produced in the original glycerides from which they were derived. Cotton-oil stearine shows a less intense color perhaps than the original oil; while in the case of tallow and lard stearines the coloration is not marked enough to be susceptible of description.

(*d*) *Refractive index.*—The refractive index of the stearines appears to be slightly lower than that of the original glycerides. The high refractive index which was noticed in the case of the original glycerides of the cotton oil was also found in the stearine from that source.

(*e*) *Rise of temperature with sulphuric acid.*—With the lard and tallow stearines no degree of comparison can be made in the rise of temperature with that produced in the original glycerides, on account of the high initial temperature which is necessary for the conduct of the experiment. Allowing for the difference in initial temperature, however, the stearines deport themselves very much as the original glycerides.

B.—CHEMICAL PROPERTIES.

(*a*) *Volatile acids.*—The amount of volatile acids in the stearines mentioned is so small as to be negligible.

(b) *Saponification equivalent.*—The numbers are essentially the same as those of the original glycerides.

(c) *Iodine number.*—The percentage of iodine absorbed by the stearines is, as is to be expected from the fact that they contain less triolein, markedly less than that of the original glycerides. The fact that the stearines possess that property in this diminished degree is of quite as much importance from an analytical point of view as their high melting point. Thus the mixture of a stearine with a low iodine number with cotton oil of a high iodine number shows a percentage of iodine absorption not greatly different from that of pure lard. One prime oleo-stearine examined showed an iodine absorption of only 17.38 per cent. Another oleo-stearine showed 26.81 per cent. The lard stearines showed higher numbers, viz, in two cases 44.24 per cent. and 49.78 per cent. The cotton-oil stearines showed iodine numbers varying from 85.28 per cent. to 99.39 per cent.

(d) *Reaction with nitrate of silver.*—The stearines react with nitrate of silver in a manner entirely comparable with that of their original glycerides. The colors, however, are not so marked nor the precipitate of silver quite so abundant with cotton-oil stearines as with the oils themselves.

(e) *Microscopical appearances.*—Stearine derived from beef or mutton tallow shows under the microscope the characteristic fan shaped crystals already noticed. Lard stearine, on the other hand, gives crystalline groups similar to those already mentioned in the case of lard.

(f) *Moisture.*—Properly prepared stearine contains only a trace of moisture.

OTHER ADULTERANTS OF LARD.

It has been claimed that other substances than those mentioned have been used in the adulteration of lard, but these claims seem to rest on no valid foundation. Among these substances, dead-hog grease or dead-hog stearine is the one most frequently mentioned. The term dead-hog grease is used to indicate the oil or lard obtained from animals which die of disease, or are smothered in transportation, or die on the way to the slaughtering houses. The fat of animals very recently dead, unless death takes place from disease, and taken before any decomposition sets in, has chemically the same characteristics as that derived from animals slaughtered. If, however, the animals have been dead some time before rendering a considerable decomposition of the glycerides takes place and the amount of free acid in the fat is thus largely increased. Such fat also shows a distinctly unpleasant odor, by which it can readily be detected from genuine lard. Peanut oil and some other vegetable oils have also been mentioned as adulterants of lard. While it may be true that many attempts have been made to use the above substances in the adulteration of lard on a small scale, it is also quite true that such attempts have never attained any importance from a commercial point of view.

(6) PROPERTIES OF ADULTERATED LARDS.

In external appearances to an unskilled person adulterated lards are not appreciably different from the pure article. An expert, however, is generally able to tell, by taste, odor, touch, and grain, a mixed lard from a pure one. There is usually enough lard in the adulterated article to give to it the taste and odor of a genuine one. Mixtures of fat, however, have been made, and perhaps sold as lard, which contained no hog grease whatever.* In the following descriptions an endeavor has been made to give the chief characteristics of an adulterated lard on the same plan as the descriptions of pure lard and the adulterations thereof which precede.

A.—PHYSICAL PROPERTIES.

(*a*) *Specific gravity.*—But little stress can be laid upon the numbers representing the specific gravity of adulterated lards since the materials of which they are composed have nearly the same specific gravity as the pure article. The addition of cotton oil, however, raises the specific gravity, and when this substance is present in quantities above 15 per cent. its influence on the specific gravity of the sample is marked. At 35° the specific gravity of adulterated lards varies from .906 to .910, compared with water at same temperature.

(*b*) *Melting point.*—The melting point of the adulterated lards is in most cases nearly the same as that of pure lards, but in some samples lower. This arises from the fact, which has already been noticed, of the low melting point of the cotton oil, which is one of the principal adulterants used. The numbers representing the melting points of adulterated lards, which will be found in the following tables, emphasize the fact which has already been noted that the lowering of the melting point is not theoretically proportional to the content of cotton oil found in the adulterated lards of commerce. In a number of samples of lards containing cotton oil from Fairbank & Co. the lowest melting point found was 31.5°, and the highest 41.9°, and the mean 38.1°. In the series of samples from Armour & Co. the lowest melting point noticed was 38.9°, and the highest 43.3°, and the mean 40.6°. The melting point of the Armour samples approaches much nearer that of pure kettle rendered lard than those received from Fairbank & Co. the latter being nearly the same as for steam lards. Although the melting point is not of itself a property of very great importance from an analytical point of view, yet its determination should never be neglected in a comprehensive analytical examination.

(*c*) *Color reaction.*—The amount of coloration shown by an adulterated

---

* *Cotolene* is a mixture of cotton oil and oleo-stearine, prepared by N. K. Fairbank & Co. It is sold under its true name and not as lard.

lard when treated with sulphuric or nitric acid, depends chiefly upon the percentage of cotton oil which it contains. Since from a commercial point of view the introduction of a small amount of cotton oil would not prove profitable, we find in the adulterated lards of commerce, as a general rule, strong color reactions. It might be possible, however, to mix with a pure lard so small a quantity of cotton oil as to render doubtful to the analyst the character of the color reaction produced. Some of the colors produced in the adulterated lards examined, as copied from the note-books, are as follows: "light brown," "pink red brown," "light yellow red," "light pink," "deep brown," "red," "deep red brown," etc. The appearance of a pinkish tint is often found in adulterated lards containing a notable portion of beef-fat stearine, although this coloration is not considered a certain indication of the presence of this substance.

(*d*) *Refractive index.*—The refractive index of the mixed lards naturally varies with the proportion of cotton oil which may be present. The greater the quantity of cotton oil the higher the refractive index. The refractive index of the Armour mixed lards is decidedly lower than that of the Fairbank samples. The following is the number representing the mean refractive index of the Armour samples at 25°, viz, 1.4634. The number representing the mean refractive index of the Fairbank samples is 1.4651. The refractive index is a much more important property in the sorting of lards than the melting point.

(*e*) *Rise of temperature with sulphuric acid.*—As is to be expected, we find here also great variation, depending on the nature and the quantity of the adulterants present. The presence of tallow stearine tends to diminish the rise of temperature with sulphuric acid, while cotton-oil has the opposite effect. As the relative proportion of these two ingredients and also the amount of pure lard varies, we may expect corresponding variation in the temperature shown on mixing the lard with sulphuric acid. In the samples of Armour's lards examined, the highest rise of temperature noticed was 58.9° and the lowest 42.1°. This latter number is almost identical with that furnished with pure lards. In Fairbank's lards the least rise of temperature noticed was 51.3° and the greatest 68.8°. These numbers show a larger proportion of cotton oil in the Fairbank than in the Armour samples. This rise of temperature as a diagnostic sign is valuable, and its determination should never be omitted.

(*f*) *Crystallization point of fatty acids.*—In Armour's lards the mean temperature of crystallization for the fat acids was found to be 39.8°. In the Fairbank lards it was 37.4°.

(*g*) *Melting point of fat acids.*—The mean melting point of the fat acids in the Armour samples was 42.8°. In the Fairbank samples it was 40.6°.

B.—CHEMICAL PROPERTIES.

(*a*) *Volatile acids.*—The remark which has been made in regard to the volatile acids of pure lards and their adulterants is also applicable for mixed lards. The amount is so minute as to be of no value from an analytical point of view.

(*b*) *Saponification equivalent.*—The numbers representing the saponification equivalent do not afford any particular indication of the kind of adulteration used. In the samples of Fairbank mixed lards examined the mean saponification equivalent found was 279.4. In the Armour samples it was 275.

(*c*) *Iodine number.*—The amount of iodine absorbed by a mixed lard gives a valuable indication of the kind of the ingredients which have been added to it. It has already been seen that the stearines, especially those derived from tallow, have a very low iodine number, while cottonseed oil has a very high one. It is therefore possible to mix these two substances together so that the resulting iodine number may be about the same as that of pure lard, viz, 60 per cent. In the samples of the Armour mixed lards examined the mixture seems to have been made in about the proportion indicated. The lowest iodine number observed in these lards was 54.11 per cent., which is decidedly less than that of normal pure lard. The highest number observed was 71.19 per cent. The other numbers were slightly above those obtained for pure lard. In the samples of mixed lards from Fairbank & Co. the iodine numbers are much higher. The lowest number observed was 78.24 and the highest 94.78 per cent.

(*d*) *Reaction with nitrate of silver.*[*]—Mixed lards containing cotton oil show a reduction of metallic silver in a greater or less degree, according to the proportion of cotton oil present. In every case where cotton oil was known to be present in a mixed lard this reaction was noticed. It would be possible, however, to put so small a portion of cotton oil into a lard as to render difficult the positive detection of it by the nitrate of silver test.

(*e*) *Microscopic appearances.*—The mixed lards, under the conditions described further on, show in the field of vision of the microscope distinct tufted crystals of the stearines which have been used as adulterants. The rhombic crystals of pure lard are also often noticed in this field.

(*f*) *Moisture in mixed lards.*—Mixed lards generally contain only a trace of water. In one instance, however, water appears to have been added as an adulterant, over 30 per cent. of it having been found. The use of water as an adulterant of lard, however, is not common.

---

[*] Later observations show that in samples kept for several months the reaction with nitrate of silver is indistinct and in some cases entirely absent.

LARD AND LARD ADULTERATIONS.

Comparison of properties of lard and compound lards.

The mean results of the analytical data are as follows:

Kind of samples.	Specific gravity.	Saponification equivalent.	Melting point of the glycerides.	Melting point of the fat acids.	Crystallizing point of fat acids.	Rise of temperature.	Percentage of iodine absorbed.	Refractive index.
			°	°	°	°		
Pure lard	.9053	289.3	40.7	43.3	39.6	41.5	62.48	1.4620
Lard of miscellaneous origin	.9067	274.4	41.7	42.9	39.6	45.7	64.34	1.4633
Prime steam lard	.9055	283.5	37.0	42.1	38.6	39.9	62.86	1.4623
Armour's lards	.9060	275.0	40.6	42.8	39.8	46.5	63.58	1.4634
Fairbank's lards	.9055	279.4	38.1	40.6	37.4	57.9	85.31	1.4651

STATISTICS OF THE LARD INDUSTRY.

It was developed in the investigations before the Committees on Agriculture of the Senate and House of Representatives that the annual production of lard in the United States is 600,000,000 pounds, of which about half is pure lard and the other half pure lard mixed with stearine and cotton oil, the "refined" or compound lard of commerce. The annual exports of lard are about 320,000,000 pounds, of which about 40 per cent. were compound or refined lard.*

According to the figures furnished by the Bureau of Statistics, the production of lard from 1877 to 1887, inclusive, was as follows:

Years.	Total.	Years.	Total.
	Pounds.		Pounds.
1886–'87	527,632,000	1881–'82	468,929,000
1885–'86	514,230,000	1880–'81	517,660,000
1884–'85	489,405,000	1879–'80	479,020,000
1883–'84	444,450,000	1878–'79	514,295,000
1882–'83	419,513,000	1877–'78	491,572,000

* Statement of Mr. G. H. Webster before House Committee on Agriculture. Report of Hearings, p. 26.

The exports from 1873 to 1888 are shown by the following numbers:

Years.	Lard exported.	Years.	Lard exported.
	Pounds.		Pounds.
1873	234,901,511	1881	325,001,686
1874	184,100,226	1882	239,904,657
1875	167,579,377	1883	273,236,610
1876	153,008,212	1884	228,165,733
1877	237,744,307	1885	301,305,105
1878	343,693,527	1886	295,083,094
1879	343,119,298	1887	324,515,224
1880	405,436,628		

If we take the percentage of cotton oil in the compound lard at 40, the total weight of oil used in manufacturing mixed lard is 120,000,000 pounds.

In addition to this, large quantities of cotton oil are used for salad dressing and culinary operations and in the manufacture of a substitute for lard, cotolene, which contains no hog grease whatever.

METHODS OF ANALYSIS EMPLOYED.

The processes employed in conducting the analytical work, the results of which follow, will now be briefly described.

METHOD OF TAKING THE SPECIFIC GRAVITY.

(*a*) *By the picnometer.*—Two kinds of specific gravity flasks have been used in the determinations of the specific gravities, as represented in Fig. 17, viz, a plain flask with a stopper having a capillary perforation and a flask carrying a stopper to which is attached a delicate thermometer. If the specific gravity is to be taken at a temperature of 100° or that of boiling water, the plain flask is preferable; if, however, it is to be taken at some temperature below that point, for instance, 40°, the flask with the thermometer is used. The manipulation in both cases is the same.

(*b*) *Graduation of the flasks.*—The flasks, having been cleaned, are rinsed with alcohol and ether and thoroughly dried, care being taken that the ether and alcohol vapors are removed from the interior of the flask. The flask, after it is cleaned, should be handled with dry fingers or with forceps. The stopper having been inserted, the dried and cleaned flask is weighed empty at the temperature of the balance room. If the flask be wiped with a silk handkerchief or towel before weighing it should be allowed to stand fifteen minutes in the balance before the final weights are taken. The flask is now filled with recently boiled distilled water which, to avoid mixing with air, has not been shaken. It is placed in a bath of distilled water in a vessel with a flat bottom. The bath should contain as much water as is possible to avoid flowing into the open neck of the flask. If the bath is to be kept at the boiling temperature the flask should be held steady by a wire attached to the edges of the vessel or by some other means. If the specific gravity is to be taken at a lower temperature than boiling water, say 40°, the flask having been filled with distilled water at a temperature below 40°, as described above, is closed with the stopper carrying the thermometer, which is pressed firmly to its place, care being taken that no air bubbles are occluded. The temperature of the bath is then raised slowly until it reaches 40° to 41°. The temperature of the bath is taken with another thermometer. The thermometer of the flask is carefully watched, especially as it approaches the required point. When the

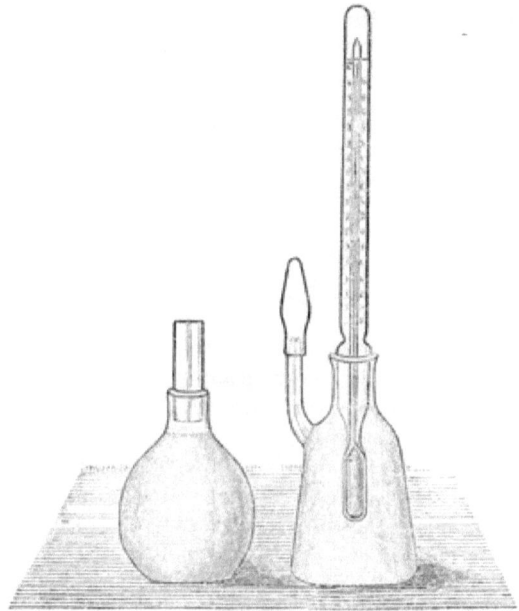

Fig. 17.

temperature of the bath is only slightly above that required the final temperature is reached only after some time, usually about one-half hour. The moment the required temperature is reached any water on top of the capillary tube is removed with blotting paper, the cap is placed upon the capillary tube and the picnometer taken from the bath; it is at once wiped perfectly dry and placed in the balance, where it is allowed to remain until the temperature indicated by the thermometer is sensibly that of the balance room; it is then weighed and the weight of distilled water which it contains at that temperature determined. When the determination is to be made at the temperature of boiling water the specific-gravity flask is secured in the bath as indicated and filled with recently boiled distilled water. The water in the bath is then brought to the boiling point by means of a lamp and the boiling continued for one-half hour. Any evaporation which may take place from the specific-gravity flask is replaced by adding a few drops of boiling distilled water. At the end of the half hour the stopper of the flask is quickly inserted and firmly pressed into its position, any water remaining on the top of the stopper being removed by a piece of filter paper. The flask is then removed from the bath, wiped perfectly dry, placed in the balance and weighed as soon as it reaches the temperature of the balance room. The weight of the distilled water which the flask contains at the given temperature having been determined, the flask is rinsed with alcohol and ether and dried as in the first instance. It is then filled with the fat, the specific gravity of which is to be determined, with the same precautions as were used in determining the weight of water.*

Example of specific gravity of fat, at 100° (boiling distilled water).

	Grams.
Weight of flask, empty	11.0956
Weight of flask+water at 100°	39.6216
Weight of water	28.5260
Weight of flask with fat at 100	36.8591
Weight of fat	25.7635
Specific gravity = 25.7635 ÷ 28.5260 = .90316.	

(c) *By the Westphal balance.*—The specific gravity of a fat can be accurately determined by a modification of the balance known as the Westphal. This instrument is shown in Fig. 18.

The principle of the apparatus may be briefly stated as follows: A glass bob is so adjusted as to be capable of displacing a given number of grames, five, for instance, of distilled water at a given temperature when wholly immersed in the liquid and suspended by a fine platinum wire. These bobs may be had graduated for any temperature, but most conveniently for those already named, viz, 35° or 40° and 100°. It is necessary for accurate work with this instrument that the temperature of the fat or oil, the specific gravity of which is to be determined, should be exactly that for which the bob is graduated, as even a slight variation from the prescribed temperature will produce a serious error in the result. In order to secure greater accuracy, especially for taking specific gravities at a temperature of 40°, a fine analytical balance can be substituted for the Westphal instrument. Such a balance arranged for use in this way is represented in Fig. 19. It is inconvenient, however,

---

*To facilitate the escape of any occluded air in placing the stoppers in the flasks, I have had the stoppers constructed with a concave bottom, the center of the concavity being at the opening of the capillary tube. The top of the stopper is also ground to a fine edge, so that any liquid that may issue from the capillary tube may flow away and thus escape absorption.

to use the ordinary balance for this method for temperatures near the boiling point on account of the difficulty of conducting the condensed vapors out of the balance case. For our work, therefore, we have used this balance only for lower temperatures.

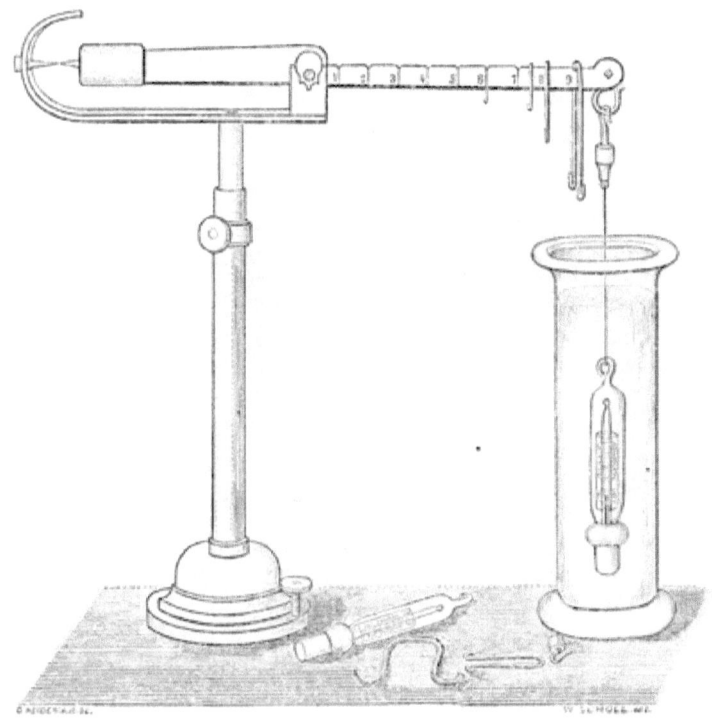

Fig. 18.

DESCRIPTION OF INSTRUMENT.

The Mohr or Westphal balance is well illustrated in the figure. The position of the instrument is shown in equilibrium. The bob is furnished with a delicate thermometer. If the bob be graduated for the displacement of exactly 5 grams of distilled water at 35°, for instance, a deep red line indicates that point. The weights are determined on the principle of the ordinary rider. There is one weight for the 5 grams and one for each 5 of decimal places of the under gram weight. The beam is so adjusted as to be in exact equilibrium when the dry bob is suspended in air. It is divided into ten parts. The big weight counts 5 when placed directly over the suspension point of the bob; 4.5 when placed at 9; 4.0 when placed at 8, etc.; when a lighter weight falls on the same figure with a heavier it is suspended from the hook of the latter.

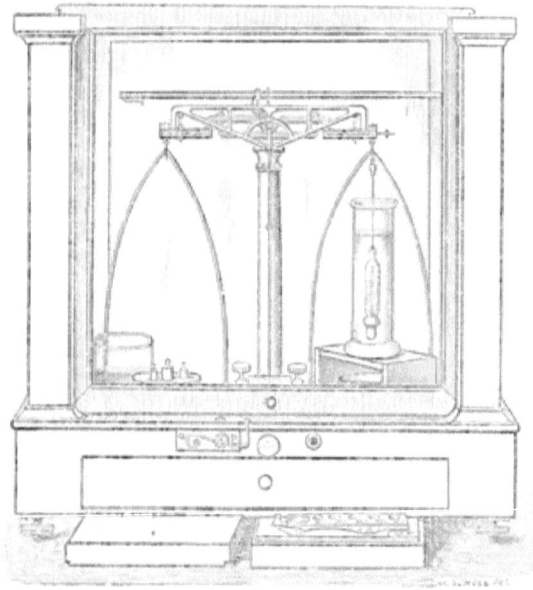

FIG. 19.

For liquids lighter than distilled water the numbers on the beam may be taken to represent the specific gravity.

Example.

Let the 5 g weight be at 9.
the .5 g weight be at 1.
the .05 g weight be at 4.
the .005 g weight be at 5.

If the beam is in equilibrium at this disposition of the weights and the temperature of the liquid that of the red mark on the bob, the specific gravity would be .9145. The actual weight of liquid displaced would be 4.5725 g, which divided by 5 = .9145.

Before beginning work with the balance the bob should be carefully graduated in pure distilled water, recently boiled and at the required temperature. Any variation in the caliber of the bob is thus determined, and any necessary correction can be introduced into the result obtained.

To change the expression of the specific gravity from direct comparison with water at any given temperature to the standard of water at 15.5° or 4°, the factor of the co-efficient of expansion of water must be introduced. One cubic centimeter of water at 35° weighs 99.418 g.

Therefore a bob which displaces 5 g of water at 35° has a volume 5.029cc. This volume of water at 4° would weigh, therefore, 5.029 g. The above specific gravity referred to water at 4° would be 4.5725÷5.029 = .9092.

In tabular form the above data are as follows:

	Grams.
Weight of 5.029cc oil at 35°	4.5725
Weight of 5.029cc water at 35°	5.0000
Weight of 5.029cc of water at 4°	5.0290
Relative weight of oil at 35° to water at 35° equals	.9145
Relative weight of oil at 35° to water at 4° equals	.9092

The change in volume of a fat or oil for each degree of temperature is approximately .0007cc for each cubic centimeter of the oil. The weight of a given volume of an oil having been determined at any temperature, its weight at the required temperature can be approximately calculated.

Thus, in the above case—

	Grams.
5.029cc of oil at 35° weighs	4.5725
Then 5.029cc of oil at 4° weighs	4.6707
Then relative weight of oil at 4° to water at 4° equals	.9307
Then relative weight of oil at 4° to water at 35° equals	.9342

ESTIMATION OF THE SPECIFIC GRAVITY OF FATS, STEARINES, ETC., IN A SOLID CONDITION.*

(a) *Estimation of specific gravity at zero.*—A platinum crucible containing about 20cc is furnished with a fine platinum bail, which is fastened through two small holes drilled into the crucible at opposite points near the upper edge. To the handle of the crucible at the central point is fastened a fine platinum wire, furnished with a loop above, by means of which it is suspended from the hook of the balance. The crucible is weighed empty and then in water at zero. This is accomplished in the following way:

The pan of the balance is protected by a wooden bench in the ordinary way in taking specific gravities, and on this bench is placed a large beaker glass containing a smaller one. The space between the two beakers is filled with finely-powdered ice and the small beaker is nearly filled with distilled ice-water. The platinum crucible is suspended at such a height as to allow it to be wholly immersed in the water, including the bail and a small portion of the suspending platinum wire. The weight of the crucible having been determined in the water, it is taken out, carefully dried, and about 15 grams of the filtered and melted fat placed in it. The fat is allowed to solidify slowly at ordinary temperatures. The crucible with fat is then weighed in the air and placed in

---

* Wollny, Milch Zeitung, 1888, No. 25 et seq.

the ice-cold water as before and weighed. Before weighing it should be allowed to stand for one hour in the water at zero.

Let t' represent the weight of the empty crucible in the air.
Let t'' represent the weight of the empty crucible in the water.
Let b' represent the weight of the filled crucible in the air.
Let b'' represent the weight of the filled crucible in the water.
Let S represent the specific gravity. Then S is computed as follows:

$$S = \frac{b' - t'}{b' - b'' - t' + t''}.$$

(*b*) In the same manner the specific gravity can be computed at 15°, 20° or 25°, or at any higher temperature at which the fat or stearine will remain in solid condition.

(*c*) *Specific gravity by Sprengel's tube.*—(For account of this method of procedure consult Allen's Commercial Organic Analysis, vol 1, page 5.)

Much confusion has arisen concerning the real meaning of the specific gravities reported for lards and lard adulterants because of failure on the part of the authors to state all the conditions. All statements of specific gravities should be accompanied by the temperature at which they were taken and the temperature of the equal volume of water with which they are compared. It would be convenient if some uniform practice of stating specific gravities could be adopted by all analysts.

(*d*) *Notes on methods of computing specific gravities.*—The rates of expansion of lard and the fat oils used as lard were carefully studied by Dr. C. A. Crampton, and I insert here his observations thereon.

All the determinations were made very carefully by the methods described, and the figures given are in all cases the average of two or more duplicates. In the densities taken at low temperatures the flasks filled with the samples were placed in a vessel containing water somewhat above the temperature at which the determination was to be made, and when it had dropped to this point they were carefully stoppered, taken out of the vessel, allowed to cool, and weighed. The determinations at high temperature were made by placing the flasks in an oil bath. The heat was raised to as high a point as was deemed safe, and at the temperatures used, 190° to 200° C., there was scarcely a darkening of the contents of the flasks, and I am convinced that no decomposition had taken place which would alter appreciably the density of the sample. From the densities taken at these two widely different temperatures the mean increase in density and the mean co-efficient of expansion was determined for each sample.

The formula used for this was the one usually given in the books:[*]

$$\delta = \frac{D_o - D_o'}{(t' - t)D_o'}$$

in which

D_o = density at the lower observed temperature.
D_o' = density at the higher observed temperature.
t = lower temperature.
t' = higher temperature.

[*] Watts Dictionary, Vol. III, p. 71.)

Although it appears to me that the formula as follows would be more correct—

$$\frac{D_o - D_o'}{(t - t') D_o}$$

or, still better:

$$\frac{D_o - D_o'}{(t' - t) \times \dfrac{D_o \times D_o'}{2}}$$

Great confusion exists in chemical literature in the expression of specific gravities. I have referred all my results to water at 4° C., believing that eventually all specific gravities will be stated in these terms, as is the custom in continental Europe.

The absolute densities are calculated from the formula $\Delta = \delta + \kappa$, in which

Δ = Co-efficient of absolute expansion.
δ = Co-efficient of apparent expansion in glass.
κ = Co-efficient of cubical expansion of glass = .000025.

The weights were not reduced to a vacuum, and no correction was made for the thread of mercury projecting above the bulb. The specific gravities of the different samples at various temperatures are also given in the the last columns of the table, these having been calculated by means of the co-efficients of expansion.

436 FOODS AND FOOD ADULTERANTS.

Specific gravities of fats and oils at various temper-

Serial number.	Description.	Temperature.	Apparent specific gravity in glass vessels.	Absolute specific gravity.	Temperature.	Apparent specific gravity in glass vessels.
	Lards.	°C.			°C.	
5674	Leaf lard rendered in laboratory U. S. Department Agriculture	+40 / +4	.89670	.89709	+100 / +4	.89475
5673	Intestinal lard rendered in laboratory U. S. Department Agriculture	+40 / +4	.89635	.89725	+100 / +4	.89315
5672	Head lard rendered in laboratory U. S. Department Agriculture	+40 / +4	.89816	.89906	+100 / +4	.89512
5501	Squires' pure lard, made by J. P. Squires & Co., Boston, Mass	+40 / +4	.89700	.89790	+97 / +4	.89719
5606	Cassard's pure lard, made by Cassard & Co., Baltimore, Md	+40 / +4	.89848	.89938	+187 / +4	.80888
5610	Armour's compound lard, made by Armour & Co., Chicago, Ill	+40 / +4	.89910	.90000	+100 / +4	.89616
5611	Armour's compound lard, made by Armour & Co., Chicago, Ill	+40 / +4	.89851	.89941	+110 / +4	.89522
5646	Fairbank's compound lard, made by Fairbank & Co., Chicago, Ill	+40 / +4	.90000	.90090	+100 / +4	.89724
	Lard stearines.					
5613	Lard stearine used in Armour's compound lard	+50 / +4	.88861	.88951	+200 / +4	.79579
5643	Lard stearine used in Fairbank's compound lard	+50 / +4	.88850	.88945	+200 / +4	.79677
	Beef fat and oleo-stearines.					
5607	Pure beef fat from the testicle, obtained from Prof. S. P. Sharpless	+50 / +4	.88908	.89117	+100 / +4	.80289
5612	Oleo-stearine, used in Armour's compound lard	+50 / +4	.88643	.88740	+200 / +4	.79487
5614	Oleo-stearine, used in Fairbank's compound lard	+50 / +4	.88725	.886-8	+200 / +4	.79386
	Cottonseed stearine.					
5675	Cottonseed stearine, obtained from Prof. D. Wesson	+40 / +4	.90343	.90433	+100 / +4	.90889
	Cottonseed oils.					
5687	Crude cottonseed oil, obtained from Prof. D. Wesson	+23 / +4	.91518	.91395	+100 / +4	.81118
5682	Crude cottonseed oil, obtained from Southern Cotton Oil Trust	+23 / +4	.91714	.91734	+100 / +4	.81523
5683	Summer yellow cottonseed oil, obtained from Southern Cotton Oil Trust	+23 / +4	.91590	.91636	+100 / +4	.81024
5684	Summer white cottonseed oil, obtained from Southern Cotton Oil Trust	+23 / +4	.91578	.91625	+100 / +4	.80939
5685	Winter yellow cottonseed oil, obtained from Southern Cotton Oil Trust	+24 / +4	.91624	.91673	+100 / +4	.80925
5686	Winter white cottonseed oil, obtained from Southern Cotton Oil Trust	+24 / +4	.91656	.91686	+100 / +4	.81047
5645	Refined cottonseed oil used in Armour's compound lard	+23 / +4	.91667	.91714	+100 / +4	.80925
	Olive oils.					
5617	Pure olive oil, obtained from Z. D. Gilman, Washington, D. C	+23 / +4	.91079	.91126	+100 / +4	.80450
5624	Pure olive oil, obtained from Prof. S. P. Sharpless	+23 / +4	.91034	.91081	+100 / +4	.80538

atures, with mean co-efficient of expansion.

Absolute specific gravity.	Range in temperature.	Mean difference in specific gravity for each degree centigrade (apparent in glass vessel).	Mean co-efficient of expansion (apparent in glass vessel).	Mean co-efficient of expansion (absolute).	Specific gravity (apparent).				Serial number.
					$d = \frac{+15.5°}{+4°}$	$d = \frac{+40°}{+4°}$	$d = \frac{+50°}{+4°}$	$d = \frac{+100°}{+4°}$	
	° C.								
.80910	+40 / +190	.0006136	.0007624	.0007874	.91181	.89679	.89985	.85097	5674
.80789	+40 / +190	.0006213	.0007736	.0007986	.91157	.89635	.89014	.85007	5673
.80977	+40 / +190	.0006203	.0007704	.0007954	.91336	.89816	.89195	.86094	5672
.85546	+40 / +95	.0006147	.0007122	.0007372	.91206	.89700	.89085	.86012	5591
.81345	+40 / +187	.0006095	.0007535	.0007785	.91311	.89648	.89238	.86191	5606
.81111	+40 / +190	.0006196	.0007683	.0007933	.91458	.89940	.89320	.86222	5610
.80957	+40 / +110	.0006221	.0007726	.0007976	.91378	.89854	.89232	.86121	5611
.81189	+40 / +100	.0006181	.0007660	.0007910	.91515	.90006	.89382	.86289	5646
.80069	+50 / +200	.0006171	.0007755	.0008005	.90965	.89454	.88836	.85750	5613
.80167	+50 / +200	.0006115	.0007675	.0007925	.90959	.89461	.88850	.85792	5643
.80754	+50 / +190	.0006221	.0007748	.0007998	.91144	.89620	.88998	.85888	5607
.79977	+50 / +200	.0006083	.0007656	.0007906	.90714	.89223	.88615	.85572	5612
.79776	+50 / +200	.0006158	.0007767	.0008017	.90647	.89138	.88533	.85444	5644
.81154	+40 / +190	.0006416	.0007951	.0008201	.91884	.90313	.89671	.86163	5675
.81583	+23 / +150	.0006245	.0007659	.0007929	.92016	.90486	.89862	.86739	5687
.82036	+23 / +190	.0009000	.0007466	.0007716	.92200	.90708	.90099	.87054	5682
.81486	+23 / +190	.0006328	.0007813	.0008063	.92063	.90514	.89860	.86716	5683
.81424	+23 / +190	.0006359	.0007858	.0008108	.92055	.90497	.89861	.86981	5681
.80900	+24 / +195	.0006685	.0008302	.0008552	.92101	.90553	.89885	.86542	5685
.81183	+25 / +150	.0006397	.0007896	.0008146	.92179	.90612	.89972	.86774	5686
.81390	+23 / +195	.0006462	.0007948	.0008198	.92150	.90573	.89930	.86714	5645
.80885	+22 / +190	.0006377	.0007928	.0008178	.91557	.80095	.89451	.86163	5617
.81003	+22 / +195	.0006285	.0007804	.0008054	.91505	.80085	.89337	.86194	5624

It will be seen that the results confirm, in the main, Allen's [*] conclusions in regard to the generally uniform rate of expansion of all fats and oils. The average increase in density in my samples would be rather lower than the figure he gives (.00064) for each degree C., as his figure was evidently calculated from the formula

$$\delta = \frac{D_o - D_o'}{(t'-t)D_o}$$

Kopp [†] gives a figure very nearly the same as mine, for the mean absolute expansion co-efficient of olive oil, viz: .0008034º.

In the lards and oils in the above table, I also determined the density by the plummet at $+35°$ C. These results, together with results calculated from the flask determinations, so as to make the figures comparable, are given in the following table:

Comparison of results with plummet and with specific-gravity flask.

	With plummet. $d = \frac{+35°}{+35°}C.$	With specific gravity flask. $d = \frac{+35°}{+35°}C.$
Lards:		
1................	.9058	.90508
2................	.9047	.90484
3................	.9066	.90652
4................	.9044	.90323
5................	.9067	.90671
6................	.9069	.90797
7................	.9068	.90667
8................	.9080	.90816
Cotton-seed oil:		
1................	.9151	.91391
2................	.9164	.91510
3................	.9134	.91377
4................	.9132	.91357
5................	.9154	.91412
6................	.9138	.91475
7................	.9133	.91452
Olive oil:		
1................	.9076	.90848
2................	.9080	.90841
Mean	.9058	.90985

This short series of comparisons adds testimony to the accuracy of the Archimedean method for taking specific gravities, and it is certainly a most rapid and convenient means to this end. I went through nearly the entire series of samples used in our lard investigation, about 150 in all, in three days. I would call attention to a slight inaccuracy in Allen's description of the method. On page 14 (Vol. II), he says: "The plummet should have a displacement of exactly 5cc (in water). This should, of course, be 5 grams, otherwise he would be comparing *volume* with *weight*, as it is the *weight* of the sample displaced which is used as the numerator of the fraction of which the *weight* of water displaced is the denominator in the expression of its specific gravity. The error would not be so great where the volume and density of water were taken as being identical at 15.5° C., but still quite an appreciable error would be introduced when the determination on the sample was made at 35° to 50° C., as on a fat, for example. There is much need for more exact mathematics in calculating specific gravities, and more uniformity in methods of expressing them among chemists is greatly to be desired.

[*] Op. cit., p. 20.   [†] Lieb. Ann., 93, p.129.

(b) *Melting point.*—The term melting point applied to a glyceride does not indicate a physical state capable of being appreciated with definiteness. As usually employed it indicates the temperature at which the fat becomes transparent; but this temperature, as is well known, varies under certain conditions chiefly dependent upon the initial temperature of the body. A more definite point, and one usually capable of being ascertained, is that where a thin disk of the fat, when freed from the attraction of gravitation and left to its own molecular forces, assumes a sensibly spherical state. The melting point given in the following analytical tables, with the exceptions to be noted, has been determined by an apparatus based on the above principle. This apparatus is described in the Journal of Analytical Chemistry, volume 1, part 1, pages 39 *et seq.*

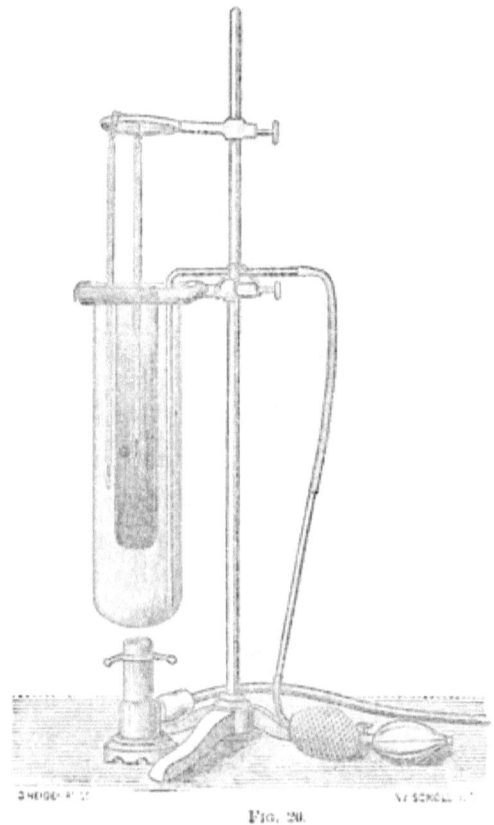

FIG. 20.

DESCRIPTION OF APPARATUS.

The apparatus, Fig. 20, consists of (1) an accurate thermometer for reading easily tenths of a degree; (2) a less accurate thermometer for measuring the temperature of water in the large beaker glass; (3) a tall beaker glass, 35cm high and 10cm in diameter; (4) a test tube 30cm high and 3.5cm in diameter; (5) a stand for supporting the apparatus; (6) some method of stirring the water in the beaker, for example, a blowing bulb of rubber and a bent glass tube extending to near the bottom of the beaker; (7) a mixture of alcohol and water of the same specific gravity as the fat to be examined.

Manipulation.—The disks of the fat are prepared as follows: The melted and filtered fat is allowed to fall from a dropping tube from a height of 15 to 20 cm on to a smooth piece of ice floating in water. The disks thus formed are from 1 to 1½ cm in diameter and weigh about 200 milligrams. By pressing the ice under the water the disks are made to float on the surface, whence they are easily removed with a steel spatula.

The mixture of alcohol and water is prepared by boiling distilled water and 95 per cent. alcohol for ten minutes to remove the gases which they may hold in solution. While still hot the water is poured into the test tube already described until it is nearly half full. The test tube is then nearly filled with the hot alcohol. It should be poured in gently down the side of the inclined tube to avoid too much mixing. If the tube is not filled until the water has cooled, the mixture will contain so many air bubbles as to be unfit for use. These bubbles will gather on the disk of fat as the temperature rises and finally force it to the top of the mixture.

The test tube containing the alcohol and water is placed in a vessel containing cold water, and the whole cooled to below 10°. The disk of fat is dropped into the tube from the spatula, and at once sinks until it reaches a part of the tube where the density of the alcohol—water is exactly equivalent to its own. Here it remains at rest and free from the action of any force save that inherent in its own molecules.

The delicate thermometer is placed in the test tube and lowered until the bulb is just above the disk. In order to secure an even temperature in all parts of the alcohol mixture in the vicinity of the disk the thermometer is moved from time to time in a circularly, pendulous manner. A tube prepared in this way will be suitable for use for several days; in fact, until the air bubbles begin to attach themselves to the disk of fat. In no case did the two liquids become so thoroughly mixed as to lose the property of holding the disk at a fixed point, even when they were kept for several weeks.

In practice, owing to the absorption of air, it has been found necessary to prepare new solutions every third or fourth day.

The disk having been placed in position, the water in the beaker glass is slowly heated and kept constantly stirred by means of the blowing apparatus already described.

When the temperature of the alcohol-water mixture rises to about 6° below the melting point the disk of fat begins to shrivel and **gradually rolls up into an irregular mass.**

The thermometer is now lowered until the fat particle is even with the center of the bulb. The bulb of the thermometer should be small, so as to indicate only the temperature of the mixture near the fat. A gentle rotary movement should be given to the thermometer bulb, which might be done with a kind of clock-work. The rise of temperature should be so regulated that the last 2° of increment require about ten minutes. The mass of fat gradually approaches the form of a sphere, and when it is sensibly so the reading of the thermometer is to be made. As soon as the temperature is taken the test tube is removed from the bath and placed again in the cooler. A second tube, containing alcohol and water, is at once placed in the bath. It is not necessary to cool the water in the bath. The test tube (ice-water being used as a cooler) is of low enough temperature to cool the bath sufficiently. After the first determination, which should be only a trial, the temperature of the bath should be so regulated as to reach a maximum about 1.5° above the melting point of the fat under examination.

Working thus with two tubes about three determinations can be made in an hour. After the test tube has been cooled the globule of fat is removed with a small spoon attached to a wire before another disk of fat is put in.

(*d*) *Refractive index.*—The apparatus used in determining the refractive index is one described by Professor Abbe in a brochure entitled "Neue Apparate zur Bestimmung des Brechungs- und Zerstreuungsvermögen fester und flüssiger Körper." The apparatus is represented in Fig. 21. For lard it is necessary that the index of refraction be de-

FIG. 21.

termined in a room where the temperature is higher than 30° and even higher than 35°. For determining the refractive index of oleo-stearines the temperature must be considerably above 40°. For very high temperatures I used the hot-room of a Turkish bath establishment. In order that the fats might quickly come to the temperature of the room and also be in a convenient apparatus for dropping upon the paper holder they were kept in a U shaped small tube-holder. The one arm of the tube was drawn out to an almost capillary diameter, bent over at the end forming a spout to facilitate the dropping of the oil upon the paper receptacle.

The apparatus is operated as follows: Fine tissue paper of rather heavy body is cut into rectangular pieces 3 cm in length by 1.5 cm in breadth. One of these pieces of paper is placed on the lower of the two glass prisms of the apparatus. Two or three drops of the oil or the fat are placed upon the paper and the upper prism carefully placed in position so as not to move the paper from its place. In charging the apparatus with the oil in this way it is placed in the horizontal position. After the paper disk holding the fat is secured by replacing the upper prism the apparatus is placed in its normal position and the index moved until the light directed through the apparatus by the mirror shows the field of vision divided into dark and light portions. The dispersion apparatus is now turned until the rainbow colors on the part between the dark and light field have disappeared. Before doing this, however, the telescope, the eye-piece of the apparatus, is so adjusted as to bring the cross-lines of the field of vision distinctly into focus. The index of the apparatus is now moved back and forth until the dark edge of the field of vision falls exactly in the intersection of the cross-lines. The refractive index of the fat under examination is then read directly upon the scale by means of a small magnifying glass. To check the accuracy of the first reading the dispersion apparatus should be turned through an angle of 180° until the colors have again disappeared and the scale of the instrument again read. These two readings should fall closely together, and their mean is the true reading of the fat under examination.

METHOD OF DETERMINING REFRACTIVE INDEX AT TEMPERATURES ABOVE THE NORMAL.

The refractive index of lards, stearines, etc., can not be taken at the ordinary room temperatures. The best method of securing the desired temperature is to place the instrument and samples in a room provided with suitable heaters for maintaining the temperature at a constant point, about 50°. Some stearines may require a slightly higher temperature. I have, in the absence of any such room in our laboratory, used the hot-rooms of the Turkish bath to good advantage. Another method suggested by Mr. Von Schweinitz has been employed. The instrument is placed on the top of an air-bath maintained at a constant temperature. The room must also be kept without change of temperature. The instrument should be allowed to remain on the bath for at least one hour before work is commenced. If necessary it can be protected with a hood, the side next the window being provided with an opening for admitting the light, and the one next the operator being entirely open.

The thermometer should rest with its bulb as closely as possible applied to the metallic casing of the prisms of the instrument. The temperature marked by it is much lower than that of the space between the prisms occupied by the film of oil under examination. For purposes of comparative readings a cotton oil is used, the refractive index of which is carefully determined at 25°.

When the temperature of the refractometer on the air bath has become constant, the same cotton oil is placed on the prisms, and after waiting for 10 to 20 minutes for the same temperature to be established, the index at that temperature is read off. The lards, stearines, etc., are then examined at that temperature and reduced to the standard of 25° by the factor determined as above. After the introduction of each fresh sample the instrument is allowed to stand for 10 to 20 minutes in order to secure a uniform temperature for all the readings.

Example.

Cotton oil, No. 6258:
R. I. at 25° = 1.4674
R. I. at N° = 1.4565
Factor, .0109
External T° = 49.5
Calculated internal T° = 57.5

At these temperatures a large number of indices of lards, stearines, fat acids, etc., was taken.

No.	Material.	Observed index.	Index corrected to 25°.
6263	Prime lard stearine	1.4505	1.4614
6262	Neutral lard	1.4505	1.4614
6264	Kettle-rendered lard	1.4500	1.4609
6260	Golden cotolene	1.4540	1.4649
6257	Oleo oil	1.4495	1.4604
6256	Oleo-stearine	1.4475	1.4584
5626	......do......	1.4470	1.4579
5681	White cotton-oil stearine	1.4550	1.4659
5680	Yellow cotton-oil stearine	1.4555	1.4664
5606	Pure lard fat acid	1.4445	1.4554
5566	Lard fat acid	1.4455	1.4564
5626	Oleo-stearine fat acid	1.4475	1.4585
5680	Cotton-oil stearine fat acid	1.4515	1.4624
5681	Cotton oil stearine	1.4475	1.4587
5576	Compound lard fat acid	1.4455	1.4564

(c) *Rise of temperature.*—The rise of temperature which fats and oils undergo when mixed with sulphuric acid was determined in the following manner. The apparatus used is represented in Fig. 22. The idea of this piece of apparatus was derived from a description given by

Dr. W. Ramsay of an apparatus used by him in the determination of the molecular weights of nitrogen trioxide and nitric peroxide.* The rise of temperature is not wholly independent of the initial tempera-

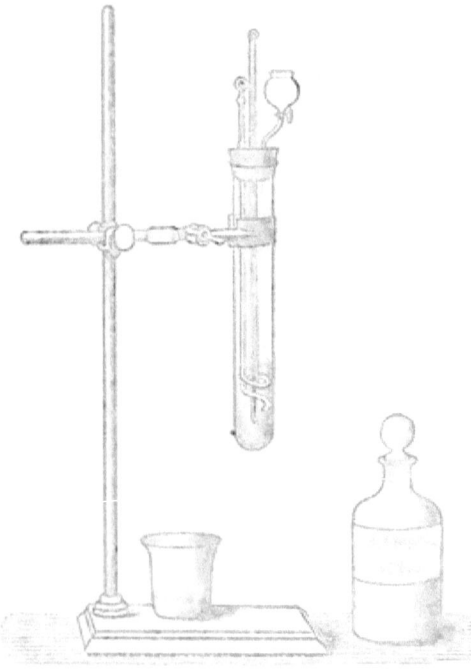

Fig. 22.

ture, and hence the initial temperature should be kept as nearly constant as possible; since most lards and adulterated lards melted at rather a high temperature are still liquid at 35° this temperature becomes a very convenient starting point. For oleo stearines the initial temperature should be 10° higher.

DESCRIPTION OF APPARATUS.

The test tube should be about 24 cm in length and 5 cm in diameter. It is furnished with a stopper with three holes, the one through the center carrying a delicate thermometer graduated to at least fifths of a degree. The second opening carries loosely a glass stirring-rod, which is bent into a coil at the lower extremity. This coil is so arranged as to have the thermometer pass through its center. The third perforation carries the funnel, which is bent outwards and upwards and holds the sulphuric acid.

* Journal of the Chemical Society, June, 1888, page 622. When in use the whole lower part of the apparatus is inclosed in a non-conducting case, as mentioned in the text

Manipulation.—Fifty cc of the fat or oil to be examined are placed in the test tube and warmed or cooled, as the case may be, until the temperature is the one required for the beginning of the experiment, say 35°; 10 cc of the strongest sulphuric acid at the same temperature are placed in the funnel, the stopper being firmly fixed in its place; the test tube containing the oil is placed in a non-conducting receptacle; the wooden cylinder lined with cork, used in sending glass bottles by mail, I have found to be convenient for this purpose. A glass rod which fits loosely in the stopper, so as to be moved rapidly up and down, is held by the right hand of the operator; with his left hand he opens the glass stopcock of the funnel and allows the sulphuric acid to flow in upon the oil. The glass stirring-rod is now moved rapidly up and down for about 20 seconds, thus securing a thorough mixture of the oil and acid. The mercury rises rapidly in the thermometer and after two or three minutes reaches a maximum, and then, after two or three minutes more, begins to descend. The reading is made at the maximum point reached by the mercury. With pure cotton oil, linseed oil, and some other substances the rise of temperature is so great as to produce ebullition in the mass, causing it to foam up and fill the tube. To avoid this smaller quantities of acid should be used or the oil in question be diluted with a less thermogenic one, so that the maximum temperature may not be high enough to produce the effect noted.

I have thought that the value of this method of work might be increased by measuring the total temperature produced in mixing given quantities of fat and sulphuric acid, and hope soon to have a calorimeter constructed suitable for this purpose.

Experiments made with the apparatus described above in regard to the influence of the initial temperature have shown that a difference of 10° in the initial temperature would cause a difference of from 2° to 3° in the maximum temperature reached during the operation.

Prof. C. E. Munroe, of Newport, has made extensive experiments on the rise of temperature produced by the mixture of oils with sulphuric acid. He has published some of the results of his work in volume 10 of the Reports of the American Public Health Association.*

In a manuscript communication from the author under date of July 20, 1888, Professor Munroe makes the following additional observations upon his method and results:

I first sought speed in mixing, using a turn-table upon which the vessel was put, or mechanical stirrer placed in the vessel or shakers, etc., but while using Maumené's proportions there were discrepancies I could not explain. So then I varied the proportion until when I reached 20 cc of oil to 25 cc of H_2SO_4 (sp. gr. 1.83) I easily got concurrent results. I had the oil and acid and vessels all at the same initial temperature. The oil was run into a beaker glass of 100 cc capacity, a delicate thermometer was inserted and initial temperature marked and compared with thermometer hanging beside burette containing oil and acid. Then acid was run in and the

---

* Volume 10, American Public Health Association, reprints for the author an article entitled "The Use of Cotton-Seed Oil as a Food and for Medicinal Purposes."

whole stirred with a glass rod, the lower part being flattened parallel to vertical axis, and stirring continued until the mercury ceased to rise.

Of course the final temperature varied with initial temperature, but the idea was to have samples of standard oils on hand and as the temperature varied in the room from day to day to make comparisons at same temperature between oil under observation and standard oils. Variations also occur with rate of stirring, but it is remarkable how close agreement is with practice. For example I cite—

Experiments made June 18, 1884, with standard oils.

Oil.	Initial temperature.	Final temperature.	Increase in temperature.	Mean.
	°	°	°	°
Lard, winter	25.4	64	38.6	
Do	25	64	39	39
Do	25	64	39	
Cottonseed, summer	26	78	52	
Do	25	79	54	53
Do	25	78	53	

Experiments with mixtures of above oils.

Oil.	Initial temperature.	Final temperature.	Increase in temperature.	Increase calculated.	Difference.	Ratio of oils taken.
	°	°	°	°	°	cc
Lard, winter	26	71.0	45	45.9	—0.9	10
Cottonseed, summer						10
Lard	26	67.4	41.4	42.5	—1.1	15
Cottonseed						5
Lard	26	73.5	47.5	49.5	—2.0	5
Cottonseed						15
Lard	26	75.5	49.5	49.5	0.0	5
Cottonseed						15
Lard	26	69.4	43.4	44.6	—1.2	12
Cottonseed						8
Lard	26	75.8	49.8	48.1	+1.7	7
Cottonseed						13
Lard	26	76.4	50.4	50.2	—0.2	4
Cottonseed						16

The **calculated** increase is based on the numbers 39° and 53° given in the first table. It will be observed that the differences reduced to percentages are large, and that the **initial** temperature of the **mixed** oils are above those of the original oils, yet there is no question about the detection of the mixture and the estimation of the proportions on a commercial scale. In point of fact what we demanded was pure lard oil only to conform to a certain standard and the test as applied secured this. I have better results than **those cited**, but this is probably as good as the average inspector will get.

It is said, however, that beech-nut-fed pork gives an **oil** that yields results like **cottonseed**. In a thorough study of the subject this should be considered as well as the effects of age and of the different methods of refining.

(*f*) *Crystallization point of fatty acids.*—The maximum temperature **reached** during the process of crystallization of the fatty acids is also a valuable indication. The method pursued in determining this point is as follows: The fatty acids were prepared in sufficient quantities to afford about 50 or 60 grams for analytical purposes. The apparatus used is represented in Fig. 23.

A very delicate thermometer with a long bulb is used, the thermometer being graduated into tenths of a degree; the readings of the mercury are made with a small eye-glass. A test tube about 15 cm in length and 2.5 to 3 cm in diameter is filled with the melted fatty acids. The temperature at which the acid is melted should be sufficiently high to secure a complete liquefaction. The tube containing the fat is placed in a stopper carried in a bottle so that the whole of the fatty acid may be contained in that part of the tube protected from external currents of air by the bottle. The bottom of this protected bottle should be warm, so that its temperature may be several degrees higher than the crystallizing point of the fatty acid. This precaution is necessary to avoid a too rapid crystallization of the fatty acid in the bottom of the test tube and to secure as nearly as possible a uniform crystallization throughout the whole mass. The thermometer is suspended in such a manner that the bulb may occupy as nearly as possible the center of the fatty mass. The thermometer should be protected from currents of air and should be kept perfectly still. The position of the mercury in the thermometer is carefully watched by the attendant as it gradually sinks toward

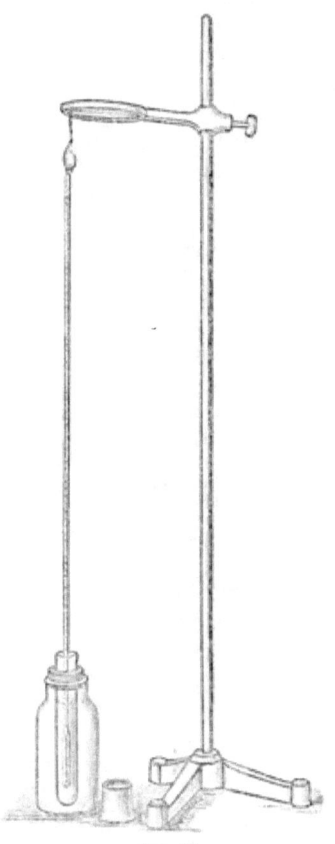

Fig. 23.

the crystallizing point. When the crystals of the acid begin to appear in the bottom and on the sides of the test tube the descent of the mercury will become very slow and finally cease; the lowest point reached by the mercury should be noted. As the crystallization extends inward toward the bulb of the thermometer a point will be reached when the mercury will begin to rise; at that point, the test tube being held by the left hand, the thermometer should be taken by the right hand of the operator and the partially crystallized mass of fat thoroughly stirred by turning the thermometer three or four times around the tube in both directions. Care should be exercised that at the end of this operation the bulb of the thermometer should hang as near as possible in the center of the crystallizing mass. Directly the above operation is accomplished the mercury will be seen to rise and this rise of temperature will continue for some time, after which the mercury will remain stationary for one or two minutes. The highest point reached is taken as the true temperature of crystallization.

(*g*) *Melting point of the fatty acids.*—The melting point of the fatty acids can not be determined in the same apparatus and by the same methods as those described for the fats themselves, because the acids are soluble in alcohol. It should be remembered that the melting point of the fatty acids is slightly above that of the glycerides, and the first determination in every case should be solely for the purpose of determining approximately the temperature at which the fat melts.

METHOD OF DETERMINING MELTING POINT OF FAT ACIDS.

(1) *By capillary tubes.*—The fat acid in a capillary tube is placed in a beaker of water, together with a delicate thermometer. The water is slowly heated and the point at which the fat becomes transparent is noted.

(2) *In a closed flask.*—This method, easy of application and giving satisfactory results, was proposed by Mr. Oma Carr.

The bulb of a delicate thermometer is coated with the fat acid, and the thermometer, by means of a cork is fastened in a round flask of 250cc capacity. The bulb of the instrument should occupy as nearly as possible the center of the flask. The cork should have an air passage for the equalization of the pressure. The flask is slowly heated in a current of warm air, or otherwise, and as the melting point is approached a rotatory movement is given to it. When the fat melts it is seen to collect in a small drop on the lowest part of the bulb, remaining stationary while the flask is turned. The thermometer is best held horizontally.

(*c*) *Color reaction.*—The re-agents used in determining the color reactions were sulphuric acid of a specific gravity 1.7 and strong nitric acid. The method of working with sulphuric acid is as follows: A porcelain plate with trough like indentations, such as is used by artists in mixing paints, is employed. The plate is warmed to a temperature slightly above that of the fat to be examined, and inclined slightly so that the liquid fat may remain in the lower end of the trough. A few drops of the fat are placed on the dish, which is capable of holding several samples; a few drops of the sulphuric acid are next placed upon the samples of fat and each one stirred with a short glass rod. The coloration produced is carefully noted, the beginning of the change in color noticed and its progress watched. The samples should also be allowed to remain for twelve hours and the coloration produced at the end of that time studied.

The method of proceeding with nitric acid is as follows:

Small test tubes are taken which are filled one-third full of the melted fat and an equal volume of the strong nitric acid is added, the test tube closed by a piece of rubber cloth, held firmly down by the thumb, and vigorously shaken for a minute. The tube is then placed in a rack and the oily layer allowed to separate from the acid. The oil being lighter rests upon the top of the acid. The coloration produced is studied in the same manner as has been indicated for sulphuric acid.

METHOD OF DETERMINING THE RELATIVE PROPORTIONS OF STEARIC AND OLEIC ACIDS IN A MIXTURE OF THE TWO.

The method proposed by Dalican and Jean rests upon the use of data of temperatures produced by the act of crystallization of the two acids. This point is determined in the manner already described.

The proportions of the two acids are then calculated from the following table:

Temperature of thermometer in degrees C.	Percentage of stearic acid.	Percentage of oleic acid.	Temperature of thermometer in degrees C.	Percentage of stearic acid.	Percentage of oleic acid.
40	35.15	59.85	45½	52.25	42.75
40½	36.10	58.90	46	53.20	41.80
41	38.00	57.00	46½	55.10	39.90
41½	38.95	56.05	47	57.95	37.05
42	39.90	55.10	47½	58.90	36.10
42½	42.75	52.25	48	61.75	33.25
43	43.70	51.30	48½	66.50	28.50
43½	44.65	50.35	49	71.25	23.75
44	47.30	47.30	49½	72.20	22.80
44½	49.40	45.60	50	75.05	19.95
45	51.30	43.70			

For mixtures of acids such as are afforded by the saponification of compound lards, the table appears to be valueless. Many of the temperatures of crystallization of such acids, as can be seen from the tables of analyses, fall below 40°.

Dr. Crampton, to whom I assigned the microscopic examination of the lards and lard compounds, has contributed the following account of the work:

CRYSTALLIZATION OF FATS.

Microscopic examination.—Probably the first application of the use of the microscope for distinguishing between fats derived from different sources was by Husson,[*] who obtained crystals from beef fat, tallow, lard, oleomargarine, goose fat, butter,

---

[*] Jour. de Pharm. et de Chim., 4ᵉ série, vol. 27, p. 100.

etc., by dissolving them in a mixture of alcohol and ether, cooling the solutions, and allowing the fats to crystallize out. He claims that he could distinguish the crystals obtained from different fats in this way, and gives illustrations made from drawings of the various forms of crystals. The delineations are very crude and poor.

In the famous "Chicago lard case"* microscopical methods were employed by experts for the defense to distinguish pure lard from lard adulterated with beef fat. Dr. W. T. Belfield seems to have been the first in that trial to point out the differences between crystals of these fats obtained from their solutions in ether or alcohol, and his methods and conclusions were followed and confirmed by most of the scientific experts employed by the defense. Dr. Belfield claimed to be able to recognize in this way as little as 10 per cent. of beef fat in lard, while some of the other gentlemen thought that as low a proportion as 5 per cent. could be shown. Twelve photomicrographs were submitted by Dr. Belfield as part of his testimony, and are reproduced in the report of the trial. They consist of crystals obtained from pure lard, both steam and kettle rendered, pure tallow, lard mixed with 30, 20, and 10 per cent. of beef tallow, and crystals from the three suspected samples of lard, "Fowler's" Nos. 1, 2, and 3. The reproduction is fairly good, but the amplification is not stated.

In Part I of this Bulletin is presented a discussion of the microscopical appearances of the various fats used in the adulteration of butter, especially the characters presented by them when viewed by polarized light, with photomicrographs prepared by Messrs. Richards & Richardson.† These are intended especially to show the use of the microscope with polarized light in distinguishing butter from its adulterants, the former having, unless it has been melted and cooled or crystallized from solvents, no crystalline structure, hence showing no refracting bodies when viewed by polarized light, while the substitutes, involving as they do in their preparation previous melting and consequent crystallization, show a variegated field. Some of these photomicrographs represent crystals obtained from lard and beef fat by crystallization from ether or alcohol, but the illumination by polarized light does not give the perfect delineation of the shape of the individual crystals necessary for their differentiation.

Examination of lards and lard substitutes with the microscope.—In the microscopical work on the larger series of samples used in the present examination, and the investigation of the efficiency of this test in distinguishing between lard and its substitutes, I had the benefit of the advice and experience of Prof. S. P. Sharpless, one of the chemists employed in the Chicago trial, who has had occasion to examine microscopically a large number of lards and lard substitutes in the course of an extensive commercial experience. Most of the samples were examined by both of us, and our results agreed with very few exceptions. I subjected the entire series of samples to a very careful examination, making several crystallizations in nearly every case, and making photo-micrographs of the appearances found in a large number of the samples, selections from which are reproduced in the plates.

Methods of procuring crystals for examination.—The methods employed by the experts in the Chicago case, in these microscopical examinations, filed as part of their testimony before the Board of Trade, are given on page.

It will be seen that there was considerable diversity in the solvents used, the manner of crystallization, and the method of preparing the crystals for examination.

My method of procedure was similar to those given, in a general way. About 2 to 5 g of the samples were taken, dissolved up in 10 to 20cc of ether in a test tube, the operation being generally hastened by warming, the tube loosely stopped with cotton, and allowed to stand over night at the ordinary temperature of the room. The proper proportions of substance and solvent can not be laid down absolutely as they are dependent upon so many conditions of temperature, solubility of the sample, etc. The proportion giving a proper rate of crystallization, neither too rapid or too slow, can

---

* McGeoch, Everingham & Co. vs. Fowler Brothers, published by Knight & Leonard, Chicago, 1883.

† Pp. 34–40.

best be found by experiment. It will differ with different samples, and with the time of year, unless the temperature of the room in which the crystallization takes place be artificially controlled.

The crystals which have formed at the bottom of the test tube are taken out with a piece of glass tubing, placed on a slide, covered with a cover glass, and examined with a ½ or ¼ inch objective. Sometimes the mother-liquor will not be sufficiently concentrated to furnish a medium for the observation of the crystals, the evaporation of the ether leaving them dry; in such cases the addition of a little cotton-seed oil will be found advantageous. I did not find any advantage in washing the crystals obtained with alcohol. I made a number of experiments in the crystallization from different solvents, alcohol, benzol, turpentine, chloroform, etc., but obtained often very different crystals from the same fat crystallized from different solvents.

On the theory that the crystals characteristic of beef fat are composed of stearine, which would probably crystallize out before the other glycerides, if present, some of the experimenters quoted above lay stress upon the examination of the first crystals formed, with the idea that these would be the beef crystals. I have not found such to be the case, the crystals formed when the solution had become concentrated being generally like those first produced, except that they were not so perfect and distinctive, having been more rapidly formed. Nor have I been able often to find such appearances as are shown by Dr. Belfield's plates of mixtures of lard with 20 and 10 per cent. of tallow, and which show the characteristic beef crystals among characteristic lard crystals on the same slide. In only two or three cases did I find the two together on the same field, and I am unable to show a single photograph of such a field, though I endeavored to make such a slide. My experience has been that the kind of crystallization first instituted predetermined the general form of all subsequent crystals.

Slight differences in the temperature or in the concentration of the solution when crystallization began seemed to have an influence upon the form of crystals produced when the substance was a mixed fat, so that in some cases where no beef crystals could be detected in a solution even by examining it at different periods, if another solution were made and allowed to crystallize, beef crystals would appear.

In Plates XXXI and XXXII are shown the characteristic crystals obtained from pure lard* when crystallized from ether; in Plate XXXVIII the crystals from beef. From these it will be seen that these fats, taken separately, give very different crystals. Just what these distinctive crystals are is a most interesting question, both in a theoretical and practical point of view. Some of the experts quoted above evidently thought, from their testimony, that the lard crystal was palmitine, and the beef stearine. Others seem to think they were both stearine modified and that this glyceride crystallizes in different forms in the different fats. Whether these different crystals are really composed of distinct glycerides, or whether they are mixtures of different but definite proportions of the various glycerides found in fat, are questions that can not be answered in the present state of our knowledge. All we can say is that they are quite different in appearance and that pure lard always gives the one, pure beef fat the other form. Not only are the forms of the individual crystals different, but the manner of aggregating themselves together is also quite distinct.

This is seen from Figs. 1 and 2, Plate XXIX, in which a small power was used in order to show the manner of aggregation of the lard crystals. They form feathery masses, radiating from a longitudinal axis, with similar secondary branches. The beef fat crystals on the other hand, as is shown in Plates XXVI and XXVIII, from spherical masses radiating from a common center, breaking up under the cover-glass into fan-

---

* The lard crystals make rather a difficult subject to photograph; they are very thin and the difference in transparency between them and the field is very little. The slight refraction of light by their edges shows their outline on the plates, but only a very delicate impression is made on account of the very thin edge.

shaped clusters, often with a peculiar twisted appearance. If the individual beef fat crystals are magnified further they still show their needle-like form, but by increasing the amplification of the cluster of lard crystals, shown in Fig. 2, Plate XXIX, for instance, we would get a similar appearance to that of Fig. 4, Plate XXX, or Fig. 9, Plate XXXIII, and by a still higher power the terminations of the crystals are plainly shown as in Fig. 5, Plate XXXI. The differences between the typical crystallization of beef and hog fat are thus easily recognized: if now the mixture of the two fats gave, on crystallizing, a mixture of the different forms in the proportion of the mixture, the recognition of such a mixed fat would be very easy even though the proportion of the one ingredient greatly preponderated. But such has not been my experience; instead of obtaining from a mixture of 10 per cent. beef fat and 90 per cent. lard, for example, a crystallization containing a great many lard crystals with a few beef fat crystals scattered amongst them, as shown by Dr. Belfield, I usually found a uniform kind of crystallization, which varied from either typical form, but which resembled more the lard. Some of these were extremely difficult to identify positively, and I was obliged to recrystallize repeatedly and vary the conditions before I could obtain sufficiently characteristic forms. Take the appearance shown in Fig. 12, Plate XXXIV, for instance: the manner of aggregation is like that of beef fat crystals, but if the individual crystals are examined by a high power, it will be seen that they are not needle-shaped and pointed, but plates with oblique terminations, although not nearly so thin or tabular as the typical lard crystals. Most of Armour's lards presented these difficulties, the appearance shown in the two figures on Plate XXXV being exceptional in this respect, and showing very plain evidence of beef admixture. Most of them gave appearances similar to Fig. 12. In two of Armour's lards, viz, serial Nos. 5557 and 5559, I was unable to find any evidence of beef fat admixture; the crystallization showing always good typical lard crystals. Compound lards from lard and cotton-seed oil only would react in this way. In Fairbank's lards on the other hand, which contain a larger proportion of beef fat, it is often difficult to obtain anything except the beef fat appearance. Pure lards sometimes give appearances, which might be mistaken for beef fat. Fig. 4, Plate XXX, for example, if viewed with a low power might possibly be mistaken for beef fat aggregations, but this mistake need not be made if the terminations of the crystals be carefully examined. The lard crystals when turned up on edge sometimes look like beef fat crystals. In the examination of a sample suspected of being compounded with beef fat, it is the beef fat appearance, of course, that is to be sought for, unless there is some special reason for knowing whether it contains any lard at all, and as soon as a characteristic beef fat crystallization is observed the object of the examination is attained. If none but lard crystals are observed at first, however, it must not be concluded at once that the sample is a pure lard, but the crystallization must be repeated, and only after a number of recrystallizations have been made, and many slides taken with no appearance of beef fat crystals can it be decided that no beef fat is present. I should say, as a result of my own observations, that as small an admixture as 20 per cent. of beef fat can readily be detected, but I should hesitate very much about guarantying a detection of 10 per cent. or less, as the experts in the Chicago case were confident of doing.

The presence of a large amount of cotton seed oil facilitates, of course, the detection of beef fat admixture by the microscope. That is, a compound lard made up with say 10 per cent. of beef fat stearine, 50 per cent. of lard, and 40 per cent. of cotton seed oil would be more likely to give a characteristic beef fat crystallization than one made up with 10 per cent. stearine, 65 per cent. lard, and 25 per cent. cotton seed oil, for the proportion of lard to beef fat would be greater in the latter case and hence more likely to predetermine a formation resembling the lard crystals. Under the ordinary conditions of crystallization no crystals would likely be obtained from the cotton seed oil. The crystallization shown in Fig. 18, Plate XXXVI, was obtained from a concentrated solution of cottonseed stearine in ether after it had stood at ordinary temperatures for nearly a week.

DESCRIPTION OF PLATES.

Plate XXIX.

Fig. 1. Lard from G. Cassard & Son, Baltimore. Serial No. 5606.
Fig. 2. Lard from intestine of hog rendered in laboratory U. S. Department Agriculture. Serial No. 5673.

Plate XXX.

Fig. 3. Lard from G. Cassard & Son, Baltimore. Different crystallization from Fig. 1. Serial No. 5606.
Fig. 4. Prime steam lard from C. H. S. Mixer, inspector, Chicago, Ill. Serial No. 5602.

Plate XXXI.

Fig. 5. Leaf lard rendered in laboratory U. S. Department Agriculture. Serial No. 5674.
Fig. 6. Lard from intestine of hog rendered in laboratory U. S. Department Agriculture. Serial No. 5673.

Plate XXXII.

Fig. 7. Lard from J. P. Squire & Co., Boston, Mass. Serial No. 5591.
Fig. 8. Same as Fig. 7. Another slide.

Plate XXXIII.

Fig. 9. Lard from Jacob Shafer, Baltimore. Serial No. 5550.
Fig. 10. Leaf lard rendered in laboratory. Serial No. 5674.

Plate XXXIV.

Fig. 11. Leaf lard rendered in laboratory. Serial No. 5674.
Fig. 12. Refined lard from Armour & Co., Chicago, Ill. Serial No. 5611.

Plate XXXV.

Fig. 13. Refined lard made by Armour & Co., Chicago, Ill. Serial No. 5610.
Fig. 14. Same as Fig. 13. Another slide.

Plate XXXVI.

Fig. 15. Refined lard made by Fairbank & Co., Chicago, Ill. Serial No. 5563.
Fig. 16. Refined lard made by Fairbank & Co., Chicago, Ill. Serial No. 5574.

Plate XXXVII.

Fig. 17. Lard stearine made by Armour & Co., Chicago, Ill. Serial No. 5613.
Fig. 18. Cotton-seed stearine from Southern Cotton Oil Trust. Serial No. 5680.

Plate XXXVIII.

Fig. 19. Oleostearine from N. K. Fairbank, Chicago, Ill. Serial No. 5644.
Fig. 20. Same as Fig. 19. Another slide.

Fig 1

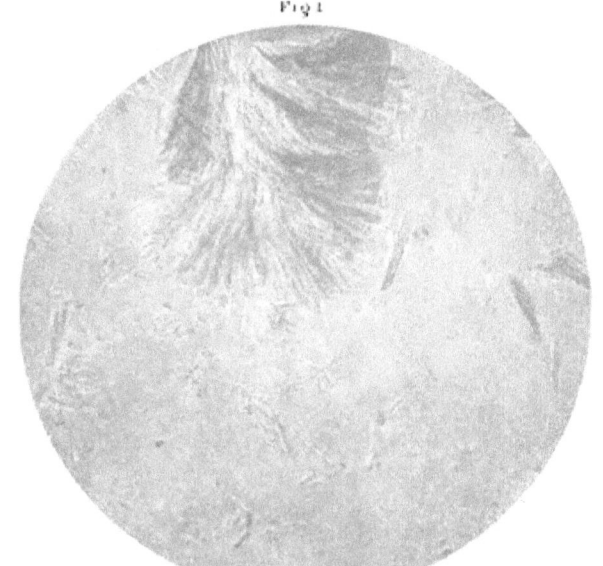

CASSARD'S LARD ×30

Fig 2

INTESTINAL LARD ×30

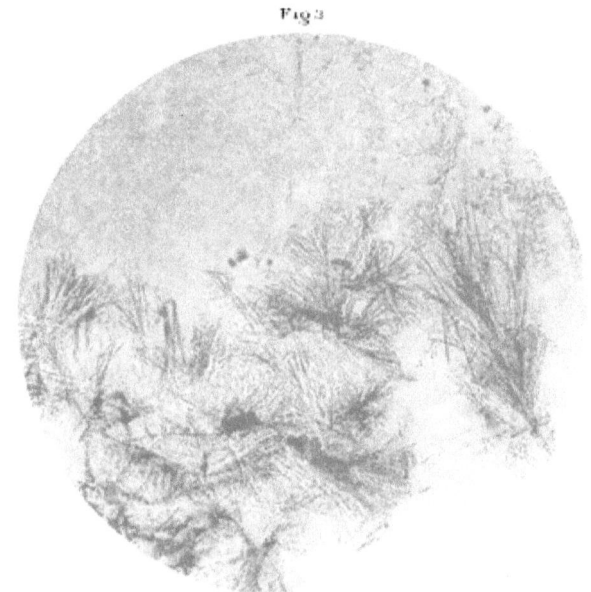

CASSARD'S LARD x 30

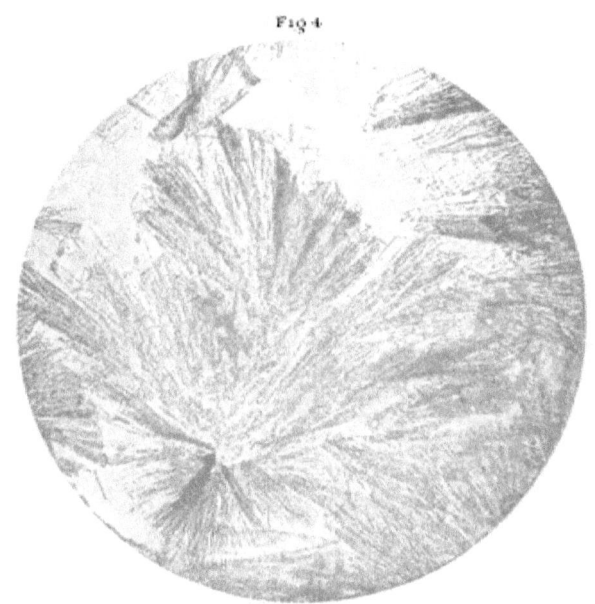

PRIME STEAM LARD x 65

LEAF LARD x185

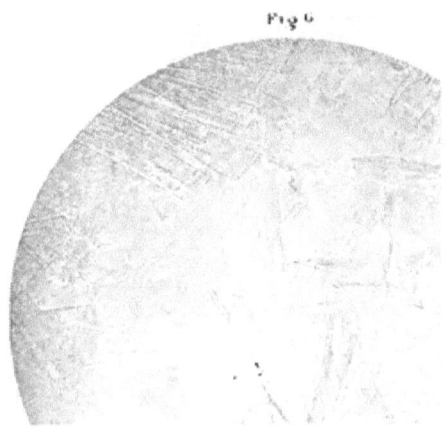

Fig 6

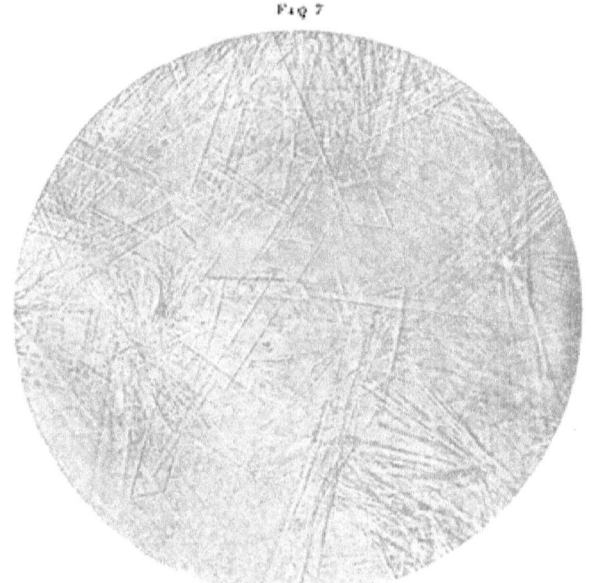

Fig 7

SQUIRE'S LARD x65

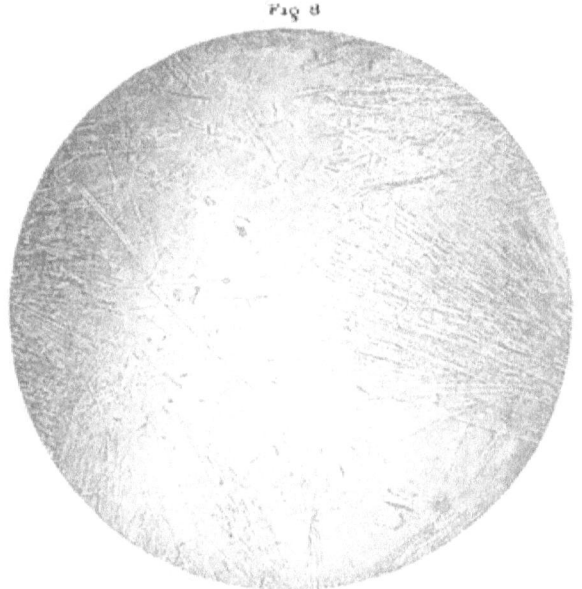

Fig 8

SQUIRE'S LARD x65

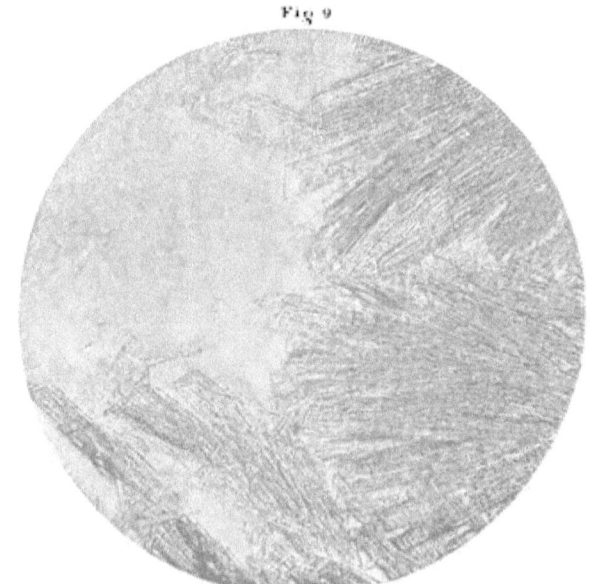

Fig 9

SHAFER'S LARD ×65

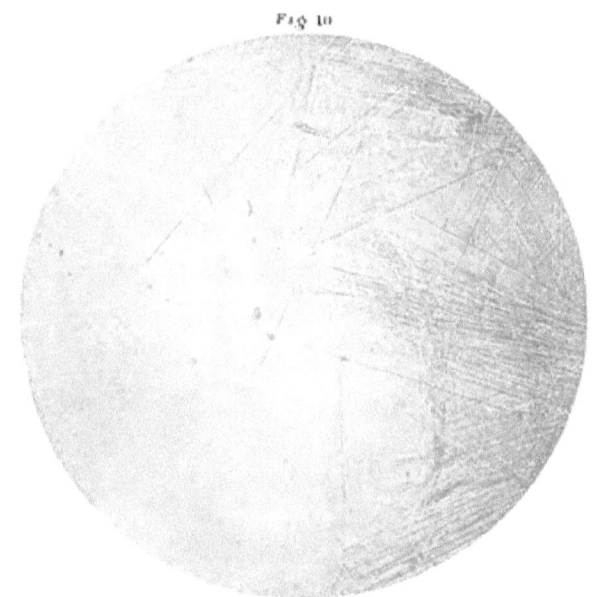

Fig 10

LEAF LARD ×65

Fig 11

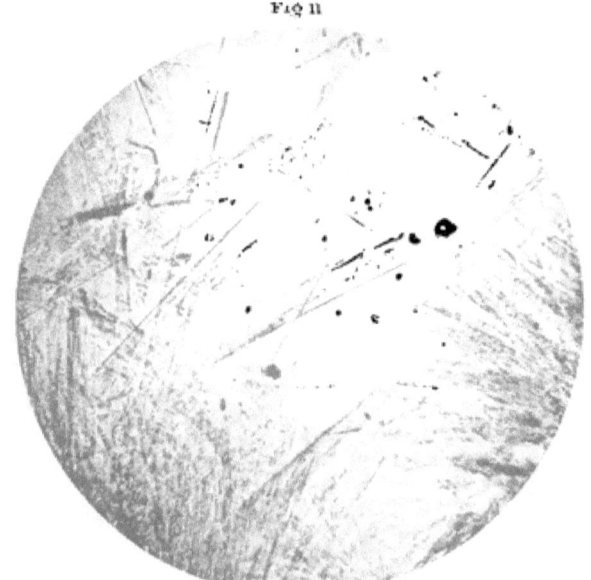

LEAF LARD x65

Fig 12

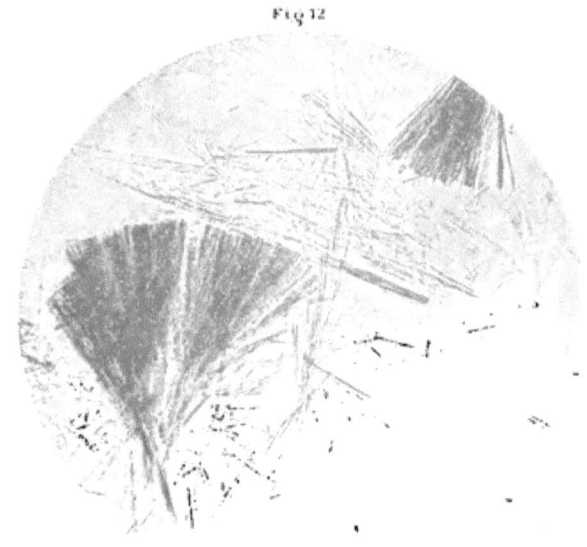

Fig 13

ARMOUR'S REFINED LARD x 65

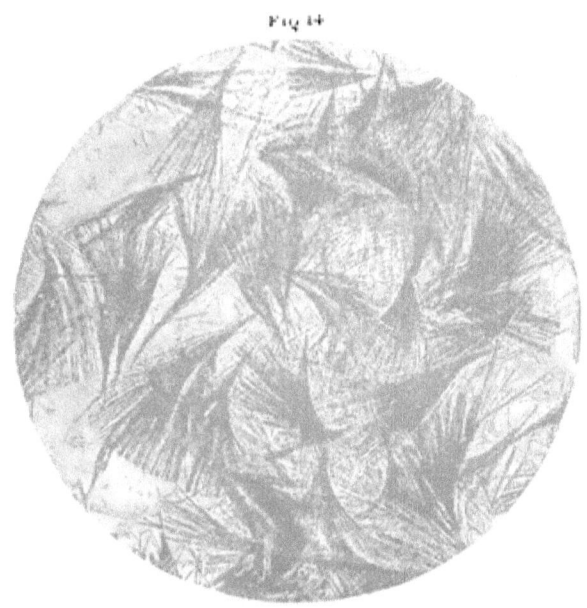

Fig 14

ARMOUR'S REFINED LARD x 65

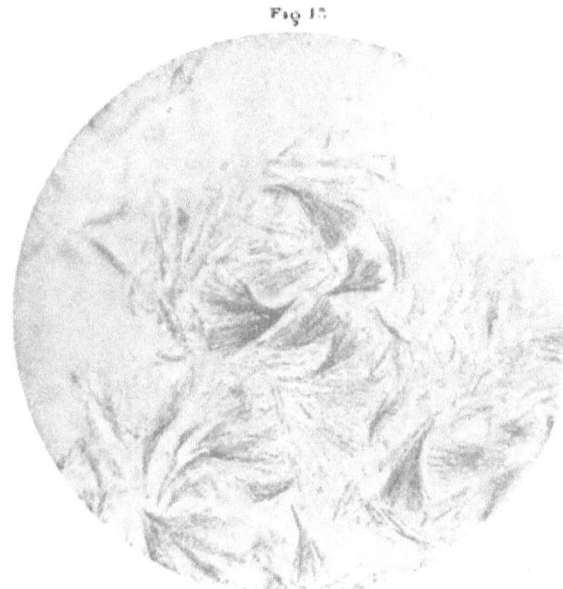

Fig 15

FAIRBANK'S REFINED LARD x 63

Fig 16

FAIRBANK'S REFINED LARD x 63

Photo. by C.A.Crampton. A.Hoen & Co. Heliocaustic. Baltimore

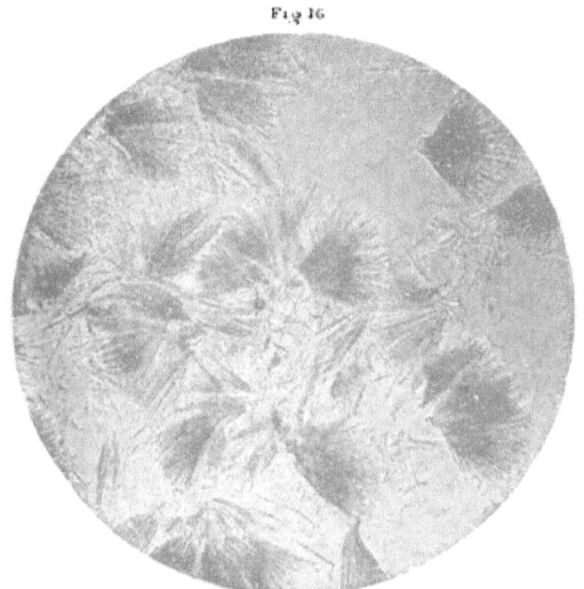

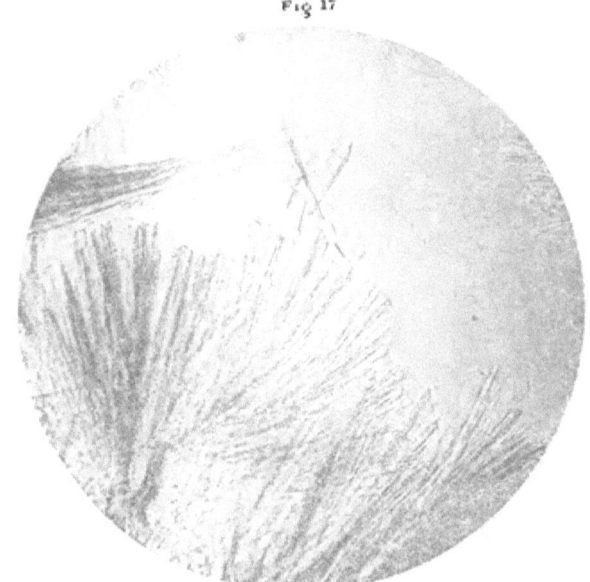

PRIME LARD STEARINE. x 65

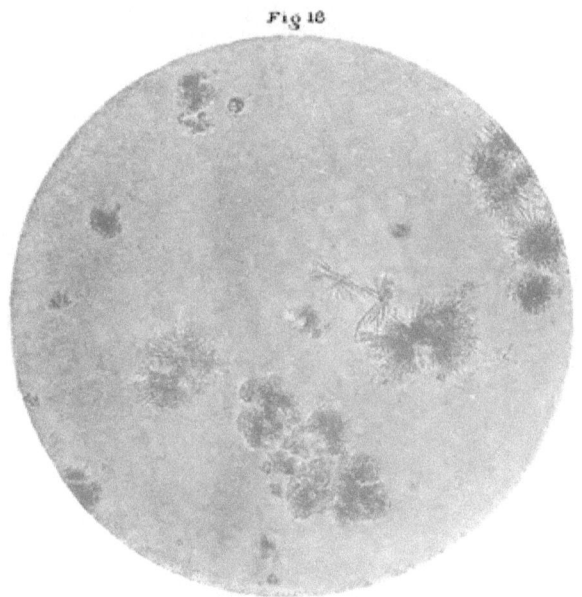

COTTON OIL STEARINE. x 65

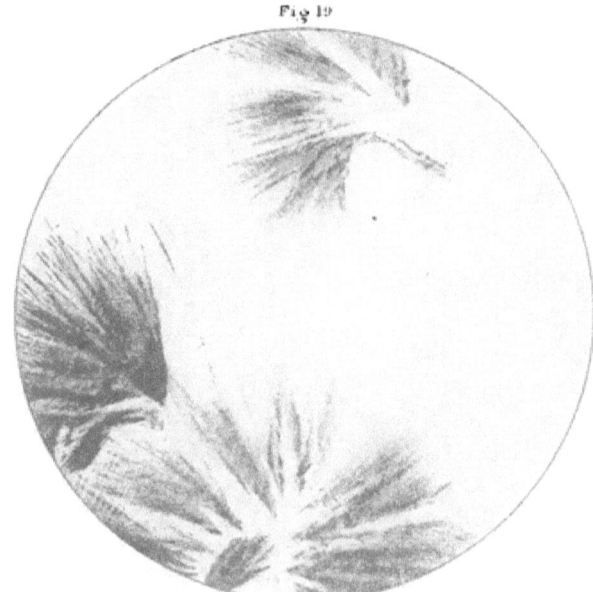

Fig 19

FAIRBANK'S OLEO-STEARINE x 65

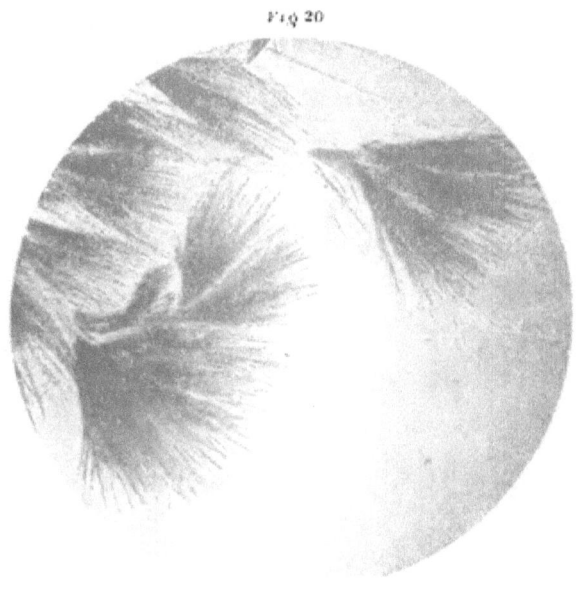

Fig 20

B.—CHEMICAL PROPERTIES.

(*a*) *Volatile or soluble and insoluble acids.*—The determination of the volatile acid is made in the apparatus represented in Fig. 24. The soluble acid may be estimated by the process described in Bulletin No. 16, a résumé of which follows.*

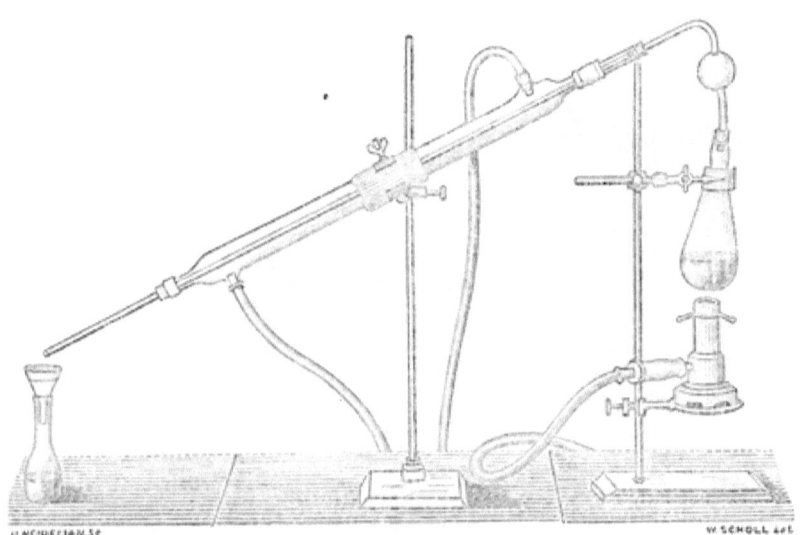

FIG. 24.

METHOD FOR THE DETERMINATION OF SOLUBLE AND INSOLUBLE ACIDS.

Reagents.—(1) A standard semi-normal hydrochloric-acid solution, accurately prepared.

(2) A standard deci-normal soda solution, accurately prepared; each 1 cc. contains .0040 grams of NaOH and neutralizes .0088 grams of butyric acid, $C_4H_8O_2$.

(3) An approximately semi-normal alcoholic potash. Dissolve 40 grams of good stick potash in 1 liter of 95 per cent. alcohol, redistilled. The solution must be clear and the KOH free from carbonates.

(4) A 1 per cent. solution of phenolphthalein in 95 per cent. alcohol.

Saponification is carried out in rubber-stoppered beer bottles holding about 250 cc, or in a round-bottom strong flask used in distillation.

About 5 grams of the melted butter fat, filtered and freed from water and salt, are weighed out by means of a small pipette and beaker, which

---
* Bulletin No. 16, page 70.

are reweighed **after the sample has been taken out and run into a saponification bottle ; 50 cc of the semi-normal potash are** added, **the** bottle closed **and placed** in the steam-bath until the contents **are** entirely saponified, facilitating the operation by occasional agitation. **The** alcoholic potash is measured always in the same pipette, and uniformity further insured by **always** allowing it to drain the same length of time, **viz, thirty** seconds. **Two or** three blanks are also measured out at the **same time** and treated **in the** same way.

In from five to thirty minutes, according to the nature of the fat, the **liquid will appear perfectly homogeneous, and** when this is the case the saponification is complete, **and the** bottle may be removed and cooled. When sufficiently cool **the stopper is removed** and the contents of the flask rinsed with a little 95 per **cent. alcohol into an** Erlenmeyer flask of about 200 **cc capacity, which is placed** on the steam-bath, together with the blanks, until the **alcohol** has evaporated.

Titrate **the** blanks with **semi**-normal HCl, using phenolphthalein as an indicator. Then **run into each of the** flasks containing the fat acids 1 cc. more **semi-normal HCl than is required** to neutralize to potash in blanks. **The flask is then connected** with a **condensing** tube 3 feet long, made **of small glass tubing, and** placed on the steam-bath until **the separated fatty acids form a clear** stratum on the surface of the liquid. **The flask and contents are then allowed to become** thoroughly **cold,** ice-water **being used for cooling.**

The fatty **acids having quite solidified, the contents of** the flask are filtered through **a dry filter paper** into a liter flask, care being taken **not to break** the cake. **Two** hundred to three hundred cubic centimeters of hot water is next **poured on** the contents of the flask, the cork with its condenser tube re-inserted, and heated on the steam-bath until the cake of acids is thoroughly melted, the flask being occasionally agitated with a circular motion, so that none of its contents are brought on the cork. When the fatty acids **have** again separated as an oily layer, the flask and its contents are cooled in ice-water, and the liquid filtered through the same filter into the same liter flask. This treatment with hot water, followed by cooling and filtration of the wash-water, is repeated three times, the **washings** being added to the first filtrate. The mixed washings and filtrate are next made up to 1 liter, and 100 cc, in duplicate, are taken and titrated with deci-normal NaOH. The volume required is calculated to the liquid. The number so obtained represents the measure of deci-normal NaOH neutralized by the soluble fatty acids of the butter fat taken, plus that corresponding to the excess of the standard acid used, viz, 1 cc. The amount of soda employed for the neutralization is to be diminished, for the 1 liter, by 5 cc, corresponding to the excess of 1 cc ½ N. acid.

This corrected volume, multiplied by the factor .0088, gives the butyric acid in the weight of butter fat employed. (See table.)

Fig 24 bis.

The flask containing the cake of insoluble fat acids is inverted and allowed to drain and dry for twelve hours (Fig. 24 bis), together with the filter paper through which its soluble fatty acids have been filtered. When dry the cake is broken up and transferred to a weighed glass evaporating dish. Remove from the dried filter paper as much of the adhering fat acids as possible and then add them to the contents of the dish. The funnel, with the filter paper, is then placed in an Erlenmeyer flask, a hole is made in the bottom of the filter paper, and it is thoroughly washed with absolute alcohol from a wash bottle. The flask is rinsed with the washings from the filter paper and pure alcohol, and these transferred to the evaporating dish. The dish is placed on the steam-bath and the alcohol driven off. It is then transferred to the air bath and dried at 100° C. for two hours, taken out, cooled in a desiccator, and weighed. It is then again placed in the air-bath and dried for another two hours, cooled as before, and weighed. If there is no considerable decrease in weight the first weight will do; otherwise, reheat two hours and weigh. This gives the weight of insoluble fat acids in the quantity taken, from which the percentage is easily calculated.

Table for the calculation of soluble fatty acids.

No. cc KOH Sol.	Equivalent.	N/10 NaOH.	Equivalent.	N/10 NaOH	Equivalent.	N/10 NaOH	Equivalent.
	Grams.		Grams.		Grams.		Grams.
10	.0880	.25	.0942	.50	.0924	.75	.0946
11	.0968	25	.0980	50	1012	75	1034
12	.1036	25	.1078	50	.1100	75	.1122
13	.1144	25	.1166	50	.1188	75	.1210
14	.1232	25	.1254	50	.1276	75	.1298
15	.1320	25	.1342	50	.1364	75	.1386
16	.1408	25	.1430	50	.1452	75	.1474
17	.1496	25	.1518	50	.1540	75	.1562
18	.1584	25	.1606	50	.1628	75	.1650
19	.1672	25	.1694	50	.1716	75	.1738
20	.1760	25	.1782	50	.1804	75	.1826
21	.1848	25	.1870	50	.1892	75	.1914
22	.1936	25	.1958	50	.1980	75	.2002
23	.2024	25	.2046	50	.2068	75	.2090
24	.2112	25	.2134	50	.2156	75	.2178
25	.2200	25	.2222	50	.2244	75	.2266
26	.2288	25	.2310	50	.2332	75	.2354
27	.2376	25	.2398	50	.2420	75	.2442
28	.2464	25	.2486	50	.2508	75	.2530
29	.2552	25	.2574	50	.2596	75	.2618
30	.2640	25	.2662	50	.2684	75	.2706

The table gives the weight of soluble acids (butyric, etc.) for each quarter of a cubic centimeter of deci-normal alkali from 10 to 30.

Example.

	Grams.
Weight fat taken	4.967
No. cc $\frac{N}{10}$ alkali used	25.50
Less 5cc due to 1cc $\frac{N}{2}$ acid	20.50
Weight soluble fat acids	.1804
Per cent. soluble fat acids	3.63

The modification introduced into the above method is in making the flask in which the saponification takes place and from which the distillation is made the same. For this purpose a specially-made flask such as is used in the digestion in the Kjeldahl method of determining nitrogen is employed. This flask is made of extra heavy glass, well annealed and quite heavy, so as to resist the pressure of the tension of the alcohol at the temperature of the steam-bath. The sample of fat with the saponifying re-agents having been placed in the flask, a stopper of soft cork is inserted and tied down with a string or wire as represented in Fig. 25. The flask is then placed upon a steam-bath and heated for one hour, at the end of which time the fats will be found saponified and any ether which may have been developed decomposed by the excess of alkali present. After cooling the stopper of the flask is removed, the alcohol evaporated, and the decomposing acid added, and the distillation carried on essentially in the manner described.

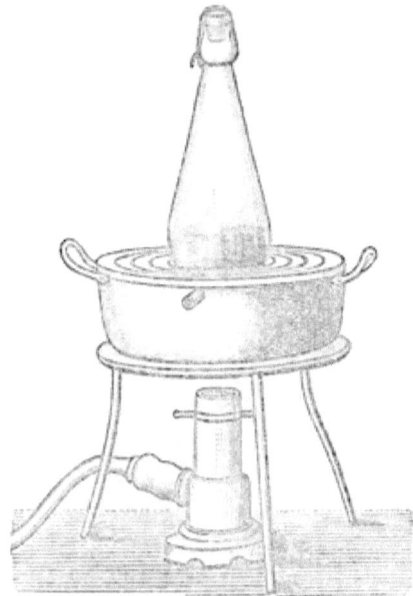

Fig. 25.

This method of procedure avoids the possibility of any loss which might ensue in transferring the saponified fats from the vessel in which the saponification took place into the distilling flask.

In evaporating the alcohol the residual soap sometimes froths and fills the flask. This is avoided by removing the flask from the steam-bath when signs of frothing are shown and rolling it in such a manner as to coat the bottom and lower fourth of the flask with a film of soap. The flask should also be inverted and waved to and fro towards the end of the evaporation in order to remove the vapor of alcohol.

The method proposed by Wollny* has also been used in the estimation of volatile acids. The method is as follows:

Five grams of the fat are weighed into an Erlenmeyer flask; 10cc of alcohol at 95 per cent. and 2cc of concentrated soda lye at 50 per cent., which has been preserved in an atmosphere free of carbonic acid, are added. The flask, furnished with a reflux condenser, is heated, with occasional shaking, in a boiling water-bath for one-quarter of an hour. The alcohol is then distilled off by allowing the flask to remain for three-quarters of an hour in a boiling-water bath. One hundred cubic centimeters of recently-boiled distilled water are then added and allowed to remain in the water until the soap is dissolved. The soap solution is then immediately decomposed with 10cc of dilute sulphuric acid (25cc sulphuric acid to 1 liter), and the flask immediately connected with the condenser. This connection is made by means of a 7mm diameter glass tube, which, 1cm above the cork, is blown into a bulb 2cm in diameter; the glass tube is carried obliquely upwards about 6cm and then bent obliquely downwards; it is connected with the condenser by a not too short rubber tube. The flask is warmed by a small flame until the insoluble acids are melted to a clear transparent liquid. The flame is then turned on with such strength that within half an hour exactly 110cc are distilled off. One hundred cubic centimeters of the distillate are filtered off, placed in a beaker glass. 1cc of phenolphthalein solution added and titrated with tenth normal barium hydrate solution when the red color is shown the contents of the beaker glass are poured back into the measuring glass in which the 100cc were measured, again poured back into the beaker, and again titrated with the barium solution until the red color becomes permanent. The distillation should take place in as nearly thirty minutes as possible.

GENERAL DIRECTIONS FOR WEIGHING THE FATS FOR THE ABOVE DETERMINATION.

The difficulty of measuring exactly 5 grams, as indicated above, is considerable. Since the specific gravity of a fat at any given temperature, say $35°$ or $40°$, is accurately known, I find it more convenient to measure out into the flask a volume of the melted fat which will weigh approximately 5 grams. This can be conveniently done by a graduated pipette, which should previously be warmed to a temperature slightly above that of the melted fat with which it is to be used. Let the specific gravity of the fat to be used at the temperature of measure-

* Milch Zeitung, No. 25, 1888.

ment be .903, then the number of cubic centimeters required to weigh 5 grams would be $5 \div .903 = 5.54$ cc. In a case of that kind, therefore, 5.6cc of the fat should be measured into the flask and its weight accurately determined.

Certain precautions are necessary in weighing the samples of fat in order to secure uniform results. Since the temperature at which the fat is manipulated must be kept approximately at from 35° to 40°, the method of weighing from a weighing-bottle is objectionable. Not only is it difficult to gauge the amount poured out from the weighing-bottle, but the falling temperature influences considerably successive weights. The samples should therefore be weighed in the flasks in which the saponification is to take place. In case this is done in an Erlenmeyer flask, it can be placed directly upon the pan of the balance. If the round-bottom flasks are used, however, they may either be held in a light beaker glass on the pan of the balance or suspended from a hook of the balance by a linen thread. The flasks in which the weighings are to be made should not be wiped with a towel or silk handkerchief within fifteen minutes of the time the weight is taken. It is best in weighing these flasks to remove the desiccating material from the interior of the balance, so as to avoid changes in the amount of moisture deposited on the sides of the flask during the time the weighing takes place. The flask should stand in or near the balance for not less than fifteen minutes before the weighing is made. The flask should be counterpoised on the weight-pan of the balance by a duplicate flask treated in the same way. The empty flask having been weighed, it is removed from the balance and a measured quantity of the fat run into it from a graduated pipette. The flask is now replaced upon the pan of the balance or suspended from a hook by a linen thread, as before described. The reweighing of the flask should not take place before five minutes, so that the fat may have time to cool.

Example.

	Grams.
Weight of flask counterbalanced	4.3611
Weight of flask plus fat	9.3711
Weight of fat	5.0100

In the above case the weight of fat which was required was 5 grams, and the amount as measured, as seen by the above results, was almost exactly that required. At the end of the operation the results can be calculated to exactly 5 grams by simple proportion.

(*b*) *Saponification equivalent.*—About 2.5 grams fat (filtered and free from water) are weighed into a patent rubber-stoppered bottle or flask, as described above, and 25cc approximately semi-normal alcoholic potash added. The exact amount taken is determined by weighing a small pipette with the beaker of fat, running the fat into the bottle from the pipette, and weighing beaker and pipette again, or the method described above may be used. The alcoholic potash is measured always in the same pipette, and uniformity further insured by always allowing it to drain the same length of time (thirty seconds). The bottle is then placed in the steam-bath, together with a blank, containing no fat. After saponification is complete and the bottles cooled, the contents are titrated with accurately semi-normal hydrochloric acid, using phenolphthalein as an indicator. The number of cubic centimeters of the acid used for the sample deducted from the number required for the blank gives the number of cubic centimeters which combines with the fat, and the saponification equivalent is calculated by the following formula, in which W equals the weight of fat taken in milligrams and N the number of cubic centimeters which have combined with the fat.

$$\text{Sap. equiv.} = \frac{2\,W}{N}$$

If it is desirable to express the number of milligrams of potash for each gram of fat employed, it can be done by dividing 5,610 by the saponification equivalent and multiplying the quotient by 10.

PRESERVATION OF THE REAGENTS USED IN DETERMINING THE VOLATILE ACIDS AND SAPONIFICATION EQUIVALENTS.

In order to secure uniformity of strength in the deci-normal and approximately semi-normal alkali solution employed in the above operations, it is necessary that they be preserved out of contact with the carbonic acid in the air. This is best done by the apparatus used for supplying burettes. In the U tube of this apparatus is placed some of the solution which is to be preserved in the flask itself. The air, therefore, which enters the bottle as the solutions are withdrawn is entirely deprived of carbonic acid by passing through the tube.

(c) *Iodine number— Reagents.*—Twenty-five grams of pure iodine dissolved in 500cc of strong alcohol. Thirty grams of mercuric chloride dissolved in 500cc of strong alcohol.

The solution of mercuric chloride is to be poured into the iodine solution. The iodine solution undergoes a constant change, by which its percentage of free iodine is diminished. This has been ascribed to the presence of impurities in the alcohol, but is doubtless due to a conversion of the iodine into hydroiodic acid and to the disturbing influence of the chloroform used in the subsequent process. There are, however, local changes in the strength of the iodine solution which are noticed from day to day, as is indicated in the examples which follow, the iodine solution being apparently stronger some days than others. These local variations may also be ascribed to the influence of the chloroform on the iodine solution. It is therefore of the utmost importance that the blank titrations in which the strength of the iodine solution is determined should be made on the measured portion of the solution, treated with the same amount of chloroform, and allowed to stand the same length of time as the samples containing the oils or fats whose iodine numbers are to be determined. By this method, although the strength of the iodine solution may appear to vary from day to day, yet this variation will take place *pari passu* with the change in the strength of the iodine solution in contact with the oils under examination. The effect therefore of this change will not be felt upon the number which expresses the percentage of iodine absorbed.

As an illustration of the progressive change in the strength of the iodine solution the following examples are given. In each case the strength was determined by titration with a deci-normal thiosulphate of soda solution:

A solution of iodine, made up as indicated, showed the following strength on the dates indicated:

Date.	Iodine solution. cc.		Thiosulphate solution. cc.
April 5, 1888	10	=	18.2
April 11, 1888	10	=	17.8
April 13, 1888	10	=	19.0
April 14, 1888	10	=	18.3
April 16, 1888	10	=	18.4
April 17, 1888	10	=	18.0
April 19, 1888	10	=	19.4
April 20, 1888	10	=	18.5
April 23, 1888	10	=	17.9

The iodine solution was now allowed to stand in a thick green glass bottle until the 10th of November, 1888. On that date it was found that 10cc of the solution of iodine required only 7.3cc of the solution of thiosulphate soda to neutralize it. It is thus seen that during that time two-thirds of the iodine had disappeared.

Deci-normal solution of thiosulphate of soda (hyposulphite of soda).—Reduce to a fine powder about 30 grams of the purest recrystallized thiosulphate of soda; spread this salt in a thin layer over a clean white blotting pad; cover with another pad and subject to pressure. After two or three minutes remove the pad, pour the powdered salt into a dish, and repeat the drying operation. It is better to put the salt all into a dish and respread therefrom on the blotting-pad than to stir the salt on the pad with a spatula. By this latter method some fibers of paper may be mixed with the salt. Weigh exactly 24.8 grams of the dried salt and make up to one liter, at the temperature at which the titrations are made, with recently-boiled distilled water. Since the pure salt is used the solution will be exactly deci-normal. Its strength, however, may be set by weighed portions of resublimated iodine, of which about 1 gram, weighed from a weighing bottle, should be taken for each determination. The solution of hyposulphite of sodium may also be set in the following way: Dissolve 3.874 grams chemically pure bichromate of potassium in distilled water and make the volume up to one liter. Place 20cc of this solution in a glass-stoppered flask to which has been added 10cc of a 10 per cent. iodide of potassium solution and 5cc of strong hydrochloric acid. Allow the thiosulphate of sodium solution to flow into the flask from a burette until the yellow color of the liquid has almost disappeared. Add a few drops of starch paste, and with constant shaking continue to add the thiosulphate solution until the blue color just disappears. The number of centimeters of thiosulphate solution used multiplied by 5 is equivalent to 1 gram of iodine.

Iodide of potassium solution.—One part of iodide of potassium in ten parts of water.

Starch paste.—One gram of starch in fine powder suspended in 100 parts of water and heated to the boiling-point. The paste must be cooled to the temperature of the room before using.

Manipulation.—The quantity of fat to be used is determined by its nature. If it consist largely of cotton oil one-half gram is sufficient. If it be mostly pure lard, one gram may be taken. A measured quantity of the fat corresponding to the weight desired is run into a recently-weighed glass-stoppered flask, and after a few minutes the weight of the flask and oil taken with precautions already noted. The fat is now dissolved in 10cc of chloroform; from 20 to 30cc of the iodine solution are then added from a burette. If the solution is not perfectly clear, more chloroform should be added. The amount of iodine employed should be large enough to leave an excess of 8 or 10cc unabsorbed at the end of the reaction. It is important, to secure comparative results, to have the amount of iodine in excess in each case approximately the same. This can be easily secured by a preliminary determination of the approximate amount of iodine absorbed. At the same time two blank determinations are made to determine the strength of the iodine solution; the manipulation in all cases being the same as in those samples containing the fat, save that the fat is omitted. After standing for two hours from 10 to 20cc of the iodide potassium solution are added and 150cc of distilled water, and the liquids thoroughly shaken together. The decinormal solution of thiosulphate of soda is added until the yellow color of the liquid has almost disappeared. The titration is continued after the addition of a few drops of starch paste until the blue color has entirely disappeared.

Example. Grams.
Weight of flask 74.1288
Weight of flask, plus fat 74.8168
Weight of fat .... .6880

After the **addition of 10cc of chloroform and 30cc** iodine solution the flask was allowed to stand **two hours.** Twenty cubic centimeters of the iodide solution **were then** added and 150cc of water, and the titration made with deci-normal thiosulphate solution, using starch paste as indicator. The amount of thiosulphate used was 21.2cc. The strength of the iodine solution determined by blank experiment was 10cc of the **iodine** solution = **19.4cc** of thiosulphate solution. Since 30cc of the **iodine** solution **were** used the amount of deci normal thiosulphate solution necessary to combine with the whole of the iodine would be 58.2cc; then 58.2cc − 21.2cc = 37cc, the number of cubic centimeters equivalent to the iodine absorbed by the fat. In a deci-normal thiosulphate solution each cubic centimeter equals .0127 **grams of** iodine; the total amount of iodine absorbed therefore was 37 × .0127 = .4699g. Then the percentage of iodine absorbed = .4699 × 100 ÷ .688 = 68.29.

To avoid the disturbing effect of the chloroform in the above process Mr. A. H. Allen recommends the use of the fat acids for absorbing iodine instead of the natural glycerides.

(d) *Reaction with nitrate of silver.*—* The solutions used have been of two kinds, viz: (a) one-fifteenth to one-tenth gram of $AgNO_3$ in 200cc of 95° alcohol and 20cc ether.

Of this solution 10cc should be taken for each test. (b) One gram $AgNO_3$ in 200cc of equal parts of alcohol and ether. Of this solution 1cc was used. The mixture of 85 parts of amyl alcohol and 15 parts of rape-seed oil was the same in both cases, 10cc of the mixture being used in each test. The method of making the test has also been changed. I use a porcelain dish 8 to 10cm in diameter. The re-agents with the oil (10cc) are thoroughly mixed by shaking in a test tube and then poured into the dish and placed on a steam bath. The contents of the dish are occasionally stirred and the heating is continued for twenty minutes. The deposition of silver on the dish is easily seen and the resulting colors show more clearly on the white porcelain.

Solution (b) acts more promptly than (a), but the results with (a) are more satisfactory.

The order of the phenomena will be found to be as follows:

For pure cotton oils.—In from two minutes to three minutes the mixture turns red. In five to ten minutes the red color becomes so brown as to appear black, in thick layers. At the end of the test metallic silver is deposited on the sides of the dish varying in color from bluish black to reddish purple. The liquid carries also particles of reduced silver and has a decided greenish tint.

With lards containing more than 20 per cent. of cotton oil the phenomena observed above are repeated, but not so promptly.

Even with very small percentages of cotton oil, the characteristic reactions are given.

Animal fats give no color, under similar treatment, or at most a faint red after twenty miuntes.

Of vegetable oils I have examined rape seed, olive, peanut, and linseed. These act with the re agent like the animal fats.

One hundred samples of lard and twenty-five samples of cotton oil have been examined by Bechi's test. In no case, where cotton-seed oil has been present, has the test failed in detecting it, except in two doubtful cases of alleged cotton oil to be mentioned further on.

Of the 100 samples of lard examined 74 were found to be adulterated with cotton oil.

SOME PECULIAR REACTIONS.

The reaction with crude cotton oil is not as sharp as with the refined oil. The deep red color of the sample seems to obscure the final color reactions.

Linseed oil gave a reddish color but no reduction of silver. In three

---

* Journal of Analytical Chemistry, Vol. 2, part 3, July, 1888.

samples of lard made by us from the leaf, guts, and head, respectively, of the same hog, the re-agent acted in the same way. There was a slightly greater coloration with the head and gut lard than with the leaf.

In samples of "prime steam lard" passed by the Chicago Board of Trade and made from the trimmings of the whole animal, not presumably including the leaf, the re-agent gave, after twenty minutes, a slight brownish red color, but no appreciable reduction.

In the whole number of examinations made there were three or four cases in which the results appeared doubtful. A slight reduction of the silver was observed and a color approaching a brown-black, but not with sufficient positiveness to prove the presence of cotton oil. These I have included under the adulterated samples.

In general, it may be said that any degree of adulteration of lard with cotton oil which would prove commercially profitable is at once detected with certainty by Bechi's test.

On the other hand very impure lards containing no cotton oil will give color reactions and a trace of reduction of metallic silver with Bechi's re-agents similar but not identical with a trace of cotton oil.

The reaction is undoubtedly the most valuable single test for cotton oil which has been proposed.

It remains to be seen what reactions lard which is made from swine fattened on mast or cotton meal will give with these re-agents. The nature of the reducing agent has not yet been determined. It has been suggested that it is an aldehyde. It appears to withstand saponification, and Milliau has lately proposed to use the test on the free, fatty acids of cotton oil.

Since cotton oil is sometimes refined with alkaline substances, and thus retains an alkaline reaction, it may happen in the application of the above test that a sufficient amount of alkali is present to reduce the nitrate of silver to oxide. In such a case the proper reaction of the cotton oil may be wholly obscured. To avoid this it is best to make the solution of nitrate of silver distinctly acid by the addition of a small quantity of pure nitric acid.

MILLIAU'S METHOD OF APPLYING THE NITRATE OF SILVER TEST.

The method of Milliau differs from that of Bechi in applying the solution of silver nitrate to the fat acids instead of to the original glycerides. The saponification may be made in any of the usual ways. About 5cc of the fat acids are sufficient for making the test, which is carried on in a test tube 12cc in length and 3cc in diameter. To the acid are added 20cc of strong alcohol and heat applied until the fat acids are dissolved. Add 2cc of a silver nitrate solution containing 30 grams of silver nitrate in 100cc of water; heat in steam bath until about one-third of the alcohol is evaporated. At the end of this time, if the samples be cotton oil or contain cotton oil the silver will be reduced to a metallic state, producing a brown or black color in the liquid, or give particles of reduced silver in the liquid or on the sides of the tube.*

I have employed the following modification of Milliau's method, which acts more satisfactorily than the original.

The re-agents are placed in a round-bottomed porcelain dish of about 50cc capacity. The silver re-agent is acidified by the addition of from .5 to 1cc of pure nitric acid. The reaction is conducted on a steam bath.

With the fat acids of cotton oil the order of phenomena is as follows:

The fat acid being thoroughly dissolved and warmed to the evaporating point of the alcohol, 2cc of silver re-agent are added and quickly stirred in with a glass rod. An almost immediate deep brown coloration is noticed, passing quickly to black. As the alcohol evaporates, the reduced silver collects in mirror-like scales and is carried onto the sides of the dish by the escaping alcohol. In a few minutes the liquid begins to grow clear again, and in ten minutes almost the whole of the reduced silver is attached to the sides of the dish.

The fatty acids for use in this modification were separated in the summer of 1888, but on account of a stress of other duties the work was not done until in December. The notes of these tests were mislaid, and in February, 1889, the work was again done.

The results of the second set of determinations were quite surprising, and lead one to suppose that the fat acids should not be kept a long while before treatment with the silver solution.†

In most cases the reactions were quick and satisfactory, but in a few cases entirely misleading. The fat acids of some cotton oils failed to give any reductions whatever, and in some of the mixed lards, where cotton-oil was known to be present, the reduction was so slight as to be wholly useless for analytical purposes. My experience with more recently-prepared samples showed that in such cases the anomalies mentioned above are not repeated. With pure lards there was also a trace of reduction noticed in some cases, which I suppose would not be seen in the freshly-prepared samples. The reaction with cottonseed oil acids, when it appeared at all was so clear and unmistakable, as to lead me to believe that in these respects the process of Milliau is an improvement on the method of Bechi. In the instances marked "trace of reduction" the separation of a slight amount of a substance was noticed, which, however, was usually of a brown color, and did not resemble in any marked degree the intense blue-black mirror-like deposit of silver by pure cotton oil. The use of the terms *trace of reduction* and *slight reduction* in the tables should not be construed into evidence of the presence of cotton-oil fat acids in the samples so marked.

The *marked* and heavy *reductions* were attended by an immediate brown color on adding the re-agent, passing rapidly into black. After heating for a few minutes the silver was deposited as black mirror-like

---

* Journal of the Chemical Society, August 31, 1888.
† This loss of reducing power in samples of cotton oil long kept has also been noticed by other observers.

particles on the sides of the dish and the liquid became almost colorless again.

WARREN'S CHLORIDE OF SULPHUR TEST.

The method of investigation employed was the one described by Warren.* It is as follows:

Five grams of the oil or mixture are weighed in a tared porcelain dish, which is well glazed both inside and out; it should have a capacity of about 4 ounces, so as to avoid loss from spitting. It should not be covered. Two cubic centimeters carbon disulphide are stirred in and 2cc of the mixture of sulphur chloride added. It is now placed on a hot water bath and well stirred until the action is fairly commenced; when solidified it is placed in a warm chamber, so as to drive off all volatile products. When two successive weighings are the same, it is ready for further operation. The mass will require breaking up, so as to allow imprisoned vapors to escape.

The color and consistency at the end of the reaction and when subsequently dried should be noticed; it is now ground up or divided as much as possible. The product may be too tough to break easily, or, if soft and sticky, a portion of the unaltered oil should be removed first.

It is transferred to a filter tube and washed with carbon disulphide, so as to remove all traces of unaltered oil, etc., which is received in a tared flask; about 200cc will suffice in any case. It is best to break up the mass after a partial exhaustion, especially when the product is hard and tough or soft and adhesive.

Oils, fats, resins, rosin oils, petroleum, etc., not acted on by sulphur chloride so as to yield solid products, may be separated. The melting-point of a fat before and after separation of the oil is an interesting and useful matter. The viscosity of a mixture containing an ingredient acted on by sulphur chloride is of importance in examining lubricating compounds. Let us, however, remember that some resins yield insoluble compounds with sulphur chloride.

It is advisable to perform the experiments in duplicate, so as to obtain a check on the result; the difference should not exceed what we allow on an ordinary commercial analysis.

The washing with disulphide is carried under pressure; a foot blower is convenient, but by closing the top of the filter tube, the clasping it with the warm hand will be sufficient. The exhaust will in some cases give a further yield of solid products; in these cases, if a larger quantity of chloride be used in the first place, a harder product will be obtained. This is not to be recommended, unless for special purposes, because uniformity is aimed at in the result, and it is not desirable to alter the oils too much.

The exhaust is weighed after removal of disulphide, and when the weighings are constant this is deducted from the contents of the dish, by which we obtain the weight of insoluble solid products. This procedure is more simple and reliable than weighing the insoluble solid product. The smell and color of the exhaust will in many cases reveal what the oil itself is, in spite of blending, refining, etc.

The color and tenacity of the solid product is so very characteristic in most cases that no difficulty will be felt in deciding what the oil or mixture is; thus arachis oil in lard or olive oil can be instantly detected from cottonseed oil. Arachis oil is largely adulterated with cotton oil, and I have no doubt that in many cases where cotton is supposed to be present as an adulterant the intention of a manufacturer has been to use arachis oil. I propose to examine this double adulteration shortly.

Sulphur chloride is sometimes decomposed when added to an oil; the deposited sulphur is removed from the exhaust by washing with ether saturated with sulphur. The oily portion is taken up, leaving the sulphur; we then obtain the weight of the exhaust minus the sulphur. If much sulphur is present the exhaust has a cloudy white appearance. This indicates, generally, that the chloride is in excess.

---

* Chem. News, March 23, 1888, p. 113.

This method evidently was suggested by the article in Watt's old dictionary (Linseed oil), in which the action of sulphur chloride on flax oil is described (incorrectly, as Mr. Warren has shown).

The method was tried by Dr. C. A. Crampton during the lard investigations on a few of the samples submitted, viz:

5645. Cotton oil.
5624. Olive oil.
5620. Peanut oil.
5556. Fairbank's lard.
5626. Tallow stearine.
5591. Squire's lard.
5672. Hog's-head lard.

On adding the re-agent, mixing thoroughly, and heating on the water-bath the oils became perfectly solid. The lards did not become solid and the stearine was not affected at all.

5591. A pure standard lard was scarcely attacked by the re-agent.

5672. Rendered in laboratory from the head fat was more readily affected by the re-agent.

Mr. Crampton made a preliminary examination of the solid products formed, but came to the conclusion that they contained no sulphur. If it be true that only oleine is attacked by the chloride of sulphur and not palmitine nor stearine, then pure lard ought to give a partial product insoluble in ether and carbon disulphide. Yet lard so treated is practically soluble in the re-agents named. The vegetable oils appear to be easily attacked by the chloride of sulphur and the action of the re-agent does not seem to wholly depend on the amount of oleine present.

I think Mr. Warren's method may prove of great value qualitatively and perhaps quantitatively.

QUANTITATIVE DETERMINATION OF ADULTERANTS IN LARD.

Many attempts have been made to determine quantitatively the amount of adulterants in lard. These attempts have not been attended with much success. They may be classified as follows:

(1) By weight of undissolved residue when the mixed fat is treated with ether.

(2) By the relative intensity of color produced by sulphuric acid and other re-agents.

(3) By the relative quantities of silver or gold reduced, or intensity of coloration in Bechi's, Millian's, and Hirschsohn's processes.

(4) By calculation from specific gravity.

(5) By calculation from iodine absorption.

(6) By calculation from refractive index.

(7) By determination of the insoluble matter produced by treatment with chloride of sulphur.

(8) By rise of temperature with sulphuric acid.

(9) By calculation from melting point.

(10) By the coloration produced by heating with nitric acid and albumen (Brullé's method).

In the peculiar conditions attending the analyses of mixed lards it is unnecessary to say that the most misleading results may be obtained by relying on any one of the above methods, and even when all are applied the real quantity of added adulterants may not be determined.

The processes indicated in Nos. 1, 2, and 3 of the foregoing classification may be dismissed without further discussion. They are entirely unreliable for any quantitative purpose.

(4) BY CALCULATION FROM SPECIFIC GRAVITY.

In the case of No. 4, approximate results could be reached were only one kind of adulterant used, the specific gravity of which, as in the case of cotton oil, is distinctly different from that of lard.

But if one adulterant be used like an oleo or lard stearine having a lower specific gravity, and another like cotton oil with a high one, the neutralizing effect of the two will render the results of the analysis unreliable.

Cotton oil, however, has a specific gravity considerably higher than that of a stearine is below the number for pure lard; hence a mixed lard containing equal portions of cotton oil and a stearine will have a higher specific gravity than pure lard. In point of fact, it may be said that where one of these adulterants is present in any notable quantity, say 15 to 30 per cent., the other is also present in proportions approximately known. It might be possible, therefore, to construct an arbitrary formula by which the disturbing effect of the second element could be allowed for. In this way some approximate number might be reached of the respective amounts of adulterants present.

Example:

Let specific gravity of pure lard at $35° = .905$
Let specific gravity of pure stearine at $35° = .903$
Let specific gravity of pure cotton oil at $35° = .913$

The theoretical specific gravity of a mixed lard composed of these bodies in the proportions stated would be as follows:

$$20 \text{ per cent. stearine} = .903 \times 20 = 18.060$$
$$30 \text{ per cent. cotton oil} = .913 \times 30 = 27.390$$
$$50 \text{ per cent. lard} = .905 \times 50 = 45.250$$
$$100 \text{ per cent.} = 90.700$$

Then theoretical specific gravity $= .907$.

It is usual to mix cotton oil and stearine in compound lards in the respective proportions mentioned above, viz, 1.5 parts to 1.*

---

* Testimony of George H. Webster, Report of Hearings before House Committee on Agriculture, p. 26.

LARD AND LARD ADULTERATIONS.   471

The specific gravity of the mixture is therefore—

   Cotton oil, 1.5 parts = .913×1.5 = 1.3695
   Stearine,   1.0 part = .903×1.0 = .9030
                          ─────   ────────
   Mixture,   2.5 parts =   2.2725

Theoretical specific gravity = .909.

The following table, therefore, will give the approximate percentage of adulterations corresponding to the specific gravities noted:

Table showing approximate percentage of adulteration corresponding to different specific gravities when the adulterants are cotton oil and stearine in respective proportions of 1.5 to 1.

Observed specific gravity, at 35°.	Pure lard.	Adulterant.	Of which there is—	
			Cotton oil.	Stearine.
	Per cent.	Per cent.	Per cent.	Per cent.
.9050	100.00	0.00	0.00	0.00
.9055	87.50	12.50	7.50	5.00
.9060	75.00	25.00	15.00	10.00
.9065	62.50	37.50	22.50	15.00
.9070	50.00	50.00	30.00	20.00
.9075	37.50	62.50	37.50	25.00
.9080	25.00	75.00	45.00	30.00
.9085	12.50	87.50	52.50	35.00
.9090	0.00	100.00	60.00	40.00

A general expression for the calculations above when applied to other standards of temperature and actual results obtained may be easily devised. The general formula however will still rest on the assumption that the cotton oil and stearine are mixed in the proportions noted, and this will be found to be practically the case.

Let s = the observed specific gravity at $t°$.
a = specific gravity of pure lard at $t°$.
b = specific gravity of pure cotton oil at $t°$.
c = specific gravity of pure stearine at $t°$.
$\dfrac{1.5 b + c}{2.5}$ = specific gravity of the mixed adulterants at $t°$.
x = per cent. of adulteration.

Then $x = \dfrac{100 (s-a)}{\dfrac{1.5b+c}{2.5} - a}$

For illustration we may apply this formula to the data collected in tables which follow. Each analyst should carefully determine for himself, in a great number of samples, the true specific gravities of various substances entering into the mixture at the temperature used by him as a standard.

Example.

Mean specific gravity of pure lard at 35°............................... .9053=a
Mean specific gravity of cotton oil at 35°.............................. .9042=b
Mean specific gravity of stearine... .9015=c
Mean specific gravity of the cotton oil and stearine adulterant (calculated). .9091=$\frac{1.5\, b+c}{2.5}$

The mean specific gravity s of Armour's was .906.
Then
$$x = \frac{100\,(.906 - .9053)}{9091 - .9053}$$

Whence

$$x = 18.42 \text{ per cent.}$$

The mean specific gravity of Fairbank's lards was .9095. This shows a theoretical adulteration of over 100 per cent., or in other words a lard composed wholly of stearines and cotton oil, in which the oil is in slightly greater proportions than those indicated above. The iodine number obtained shows that the lard approximates such a composition.

(5) BY CALCULATION FROM IODINE ABSORPTION.

The determination of the percentage of iodine absorbed by a mixed lard taken alone can not lead to any just idea of the amount of adulterant added.

In the case of specific gravities the numbers for oleo-stearine and lard stearine are near together, viz, for 35° .909 and .902, respectively. But for iodine numbers the difference is very great. In the three samples of oleo-stearine examined the mean iodine number is 20.73 per cent. In the two samples of lard stearine analyzed it is 47.02 per cent. The mean number for cotton oils is 109.02 per cent., for lard, 62.48 per cent., and for prime steam lard, 62.86 per cent. In a mixture we may find all of these ingredients, and therefore the iodine number of such a mixture may approximate that of a pure lard.

When the iodine number of a supposed adulterated lard goes above 65 per cent. there are grave reasons for suspecting an adulteration with cotton oil, but a pure lard made from certain parts of the hog may show even a higher number.

If the microscopic examination show the presence of oleo-stearine, and cotton oil be revealed by the silver or gold tests, the complexity of the problem is less confusing. The iodine number may then reveal the approximate quantities of the two adulterants present.

For example:

```
1.5 parts of cotton oil at 109 = 163.5
1.  part of oleo-stearine at 20 =  20.0
                                  ------
2.5 parts                       = 183.5
1.  part                        =  73.4 per cent.
```

Now, a mixed lard whose iodine equivalent is about 64 per cent. (Armour's) can not be made of any considerable quantity of the above mixture and pure lard. It must contain a notable quantity of lard stearine. For example:

```
40 parts of cotton oil and oleo-stearine at 74 = 2960
30 parts of lard at                        62 = 1860
30 parts of lard stearine at               47 = 1410
-----                                      ------
100 parts                                     = 6230
```

The theoretical iodine number of such a compound lard is therefore 62.30 per cent. The above hypothetical example, in the light of the analyses made, shows approximately the composition of a compound lard whose iodine number is not above 63 per cent.

In the Fairbank samples the mean iodine number is 85.31 per cent. The microscope revealed also the presence of oleo-stearine in these samples. They were presumably composed of cotton oil, lard, and oleo-stearines, and perhaps some lard. As was shown by the specific gravity they contained an excess of cotton oil. These mixtures may be represented by the following proportions:

```
10 parts oleo-stearine at 20 =  200
25 parts lard stearine at 47 = 1175
65 parts cotton oil at   109 = 7085
-----                         ------
100 parts                    = 8460
Theoretical iodine number = 84.60 per cent.
```

No formulæ can be given for computing the proportions of ingredients from the quantity of iodine absorbed, except in the tentative way indicated above, but the value of the iodine number, when thus studied with other quantitative data, is sufficiently illustrated.

(6) *By calculation from the refractive index.*—Some valuable information concerning the quantitative composition of a mixed lard may be derived from a study of the refractive index.

The mean refractive index at 25° of the samples of lard examined is 1.4620; water at the same temperature showing 1.3300; for cotton oil the number is 1.4674; for oleo stearine, 1.4582; for lard stearine, 1.4594.

The determination of a much larger number of samples of the stearines would be desirable before deciding on a permanent standard, but the above numbers will serve provisionally.

	Points.
Difference between lard and cotton oil	+54
Difference between lard and oleo stearine	−38
Difference between lard and lard stearine	−24

It thus appears that the addition of cotton oil to a lard would raise its refractive index, while the addition of the stearines would lower it.

In general it appears that two parts of stearine would neutralize the effect of one part of cotton oil. A mixture of 1.5 parts of cotton oil and 1 part of mixed stearines would have the following theroretical index:

$$1.5 \text{ parts cotton oil at } 1.4674 = 2.2011$$
$$1 \text{ part stearines at } 1.4588 = 1.4588$$
$$\overline{2.5 \text{ parts}} \qquad\qquad = 3.6599$$
$$1 \text{ part} \qquad\qquad = 1.4640$$

For a lard adulterated with the above-mixed adulterant we may use the following formulæ:

Let r = observed index at 25°
a = index at 25° of lard.
b = index at 25° of cotton oil.
c = index at 25° of stearine.
$\frac{1.5b+c}{2.5}$ = index at 25° of the mixed cotton oil and stearine.
x = per cent. of adulteration.

Then

$$x = \frac{100\,(r-a)}{\frac{1.5\,b+c}{2.5} - a}.$$

As an illustration of this formula take the mean numbers obtained in the tables of samples for lard, cotton oil, stearines, and Armour's mixtures:

Mean index of Armour's samples $r = 1.4634$
Mean index of pure lards $a = 1.4620$
Mean index of cotton oils $b = 1.4674$
Mean index of stearines $c = 1.4588$
Value of $\frac{1.5\,b+c}{2.5} = 1.4640$

Then $x = .14 \div .0020 = 70$ per cent.

According to this formula Armour's samples would have only 30 **per cent.** of pure lard, a result which is contradicted by other data. I am inclined to believe that the examination of a larger number of samples of stearine may show a higher index and thus bring the results obtained by the application of the above formula more into harmony with the other data.

The index for the Fairbank samples, 1.4651, shows that in these mixtures cotton oil has been used in greater proportions than indicated above, thus corroborating the results obtained by the other methods of analysis. Judged by the index of refraction alone, on the assumption that this index for the stearines is not much different from that of lard, the composition of a mixed lard is probably as truly indicated as by any other single method.

(7) *By determination of product formed by chloride of sulphur.*—Warren, in the articles already cited, has obtained some interesting results, and our own work has shown that much may be expected of a careful study of this process. Lack of time has prevented a full investigation and this will be made subsequently.

(8) *Rise of temperature with sulphuric acid.*—Valuable information relating to the composition of a mixed lard may be obtained by a study of rise of temperature of a given volume thereof when mixed with a definite quantity of strong sulphuric acid. The data obtained in our analyses are as follows:

Rise of temperature with—	Degrees.
Lard	41.5
Cotton oil	85.4
Oleo stearine	20.8
Lard stearine	37.7
Mean rise of temperature with the stearines	29.3

When the microscope reveals oleo stearine we may take the last number to represent the mean increment of temperature. For an adulterant composed of 1.5 parts of cotton oil and 1 part of stearine the mean rise of temperature would be 63°.

The apparent composition of a mixed lard on the above character of the adulterant would be illustrated by the following formula:

Let t = observed rise of temperature for sample.
a = rise of temperature for lard.
b = rise of temperature for cotton oil.
c = rise of temperature for stearine.

$\frac{1.5\,b+c}{2.5}$ = theoretical rise of temperature for the adulterant.

x = percentage of adulteration.

Then

$$x = \frac{100\,(t-a)}{\frac{1.5\,b+c}{2.5} - a}$$

This formula applied to the mean rise of temperature observed in Armour's samples gives the following result:

$$x = 23.3 \text{ per cent.}$$

Applied to Fairbank's samples it shows an adulteration of 76.3 per cent.

9. *Calculation from the melting point.*—The melting point of a fat is often of great value in helping to a correct understanding of its composition, but little reliance can be placed on it for quantitative purposes.

The different glycerides when mixed do not have a melting point which corresponds to the one theoretically calculated. For this reason equal mixtures of cotton seed oil and lard, instead of having a melting point of about 20°, really melt only at a much higher temperature. While, therefore, the determination of the melting point of a compound lard should not be omitted, it does not afford a basis for any reliable estimation of a quantitative nature.

10. *Heating with nitric acid and albumen.*—The coloration produced by heating the fat or oil under examination with nitric acid and albumen has also been proposed as a quantitative test. Although I have not tried this method quantitatively, I am of the opinion that it will be found of no greater value than the other color reactions already noted.

The Brullé test appears to be unaffected by free acid or rancidity, in which point it possesses an advantage over chloride of gold and in some cases over nitrate of silver.

SUMMARY.

From the methods already worked out as applied to the two classes of mixed lards examined the following general results are deducible, viz:

Method of examination by—	Sample from—	Per cent. of adulteration.
Specific gravity	Armour & Co	18.42
Refractive index	do	70.00
Rise of temperature	do	23.50
Specific gravity	Fairbank & Co	100.00
Refractive index	do	100.00
Rise of temperature	do	76.30

The mean percentage of adulteration for the Armour samples is 37.24. For the Fairbank samples it is 92.10.

It is not unusual to omit the percentage of lard stearine used in accounts given by manufacturers of the extent of adulteration. If we allow that one-third of the total adulterant is lard stearine the percentages of foreign fats in the Armour and Fairbank lards are 24.83 and 61.40 respectively.

In the foregoing discussion it has been assumed that the mean properties of a mixture of various glycerides are proportional to the quantities of each present. In the case of the melting point, we know that this is not the case, and the consideration of the melting point therefore as a factor in quantitative determinations has been omitted. It may be true that other properties are also unequally developed in a mixture, and this would add still another complication to the problem.

In the present state of our knowledge the chemist is unable to express definitely the degree of adulteration which a sample of lard has suffered. He can state with confidence whether or not a given sample is adulterated, and in the comparison of two widely different sets of samples—such as were obtained from Armour & Co. and Fairbank & Co.—he may safely say that one is adulterated to a greater degree than the other. Further than this the present state of our knowledge will not permit us to go.

RESULTS OF ANALYSES.

The samples of lards and lard compounds, whose analyses follow, were furnished by different persons, each sample usually accompanied by an affidavit showing where it was bought, name of sample, etc., or were purchased in open market by agents of the chemical division or rendered in the laboratory.

The classification was made as follows:

(1) Lards known or believed to be pure hog grease.
(2) Prime steam lards from Chicago Board of Trade.
(3) Lards of miscellaneous origin, both pure and adulterated.
(4) Cotton oils from different localities.
(5) Crude cotton oils and foots.
(6) Oleo, lard, and cotton-oil stearines.
(7) Mixed lards from Armour & Co., Chicago, Ill.
(8) Mixed lards from N. K. Fairbank & Co., Chicago, Ill.
(9) Miscellaneous oils.

Each sample is indicated by a number, and with each table is a list of these numbers, with a full description of the name of the sample, and place or person from whom obtained.

In the foregoing pages the analytical data obtained have been rather fully discussed, and only such explanatory items will be added here as may help to elucidate the tables. The amount of analytical work which has been done, as will be seen, is very large. While absolute accuracy has not been obtained, it is believed the data in general may be accepted as reliable.

Such an amount of work can only be accomplished by the united labors of several participators, and this, of course, magnifies the personal error to a certain degree.

SAMPLE OF AFFIDAVITS.

Many of the samples were accompanied by affidavits, which it is not necessary to reproduce in full. Their general tenor may be seen from the four following:

5646.

STATE OF NEW YORK,
 City and County of New York, ss:

Carl Dreier, being duly sworn, says: That he is a manager with N. K. Fairbank & Co., of Chicago; that it is the custom of the said firm to keep a stock of their refined lard on hand at their manufactory in Chicago for sale to the domestic and foreign trade; that the accompanying package of refined lard in the original can marked "H" was taken from the stock of the said firm on hand at the said manufactory, and is the same as that sold regularly by the said firm in the domestic and foreign markets as refined lard; that the said package was taken at random from the said stock without special selection, and that all of said stock is alike as to composition and quality as that usually and regularly sold in the domestic and foreign markets by the said firm and known as refined lard; that the said package was immediately fastened up, sealed, and marked "H" by this deponent, and that no ingredient or thing has been added to or extracted from the same; and deponent further says that the said package is a true, genuine, and fair sample of the refined lard as regularly made and sold by the said firm, and that it was not specially prepared for testing, analysis, or exhibition purposes, and that it is in every sense a genuine sample of refined lard.

CARL DREIER.

Sworn to before me this 16th day of February, 1888.

ALFRED JUNTZEH,
Notary Public, New York County.

[Affidavit to accompany Nos. 5550, 5551, and 5552.]

I, Walter L. Hill, on oath affirm and declare that I purchased on the seventh day of February, A. D. 1888, in the city of Washington, D. C., in the open market, the following packages of pail lard and paid for the same the price set against the respective items; that the names of the parties in whose places of business the same were purchased were as follows, to wit:

One (1) three (3) pound pail of lard, marked "Jacob Schaefer, & Co., Baltimore. Pure natural lard, bought of Henry W. Kem & Co., No. 12 Centre Market, Washington, D. C., price thirty-five (35) cents."

One (1) three (3) pound pail of lard, marked "Chas. G. Kriel, cream leaf lard, Baltimore, price thirty (30) cents;" bought of Jas. Schneider, No. 529 Centre Market, Washington, D. C.

One (1) three (3) pound pail of lard, marked "Armour & Co., pure refined family lard, Chicago," bought of E. C. Ford & Son, No. 609 Centre Market, Washington, D. C.; price thirty-five (35) cents.

And I, the said Hill, further on oath affirm and declare, that I likewise purchased on the eighth day of February, A. D. 1888, in the said city of Washington, in the open market, the following package of pail lard and paid for the same the price set against the said item, and that the name of the party in whose place of business the same was purchased is as follows, to wit:

One (1) five (5) pound pail of lard, marked "G. Cassard & Son, best refined lard, Baltimore," with a star, viz., ✯ bought of Hume, Cleary & Co., No. 807 Pennsylvania avenue, Washington, D. C.; price fifty-five (55) cents.

That I took said samples and delivered them to Professor S. F. Sharpless in the original packages, as purchased, and that the same were not in way tampered with by me.

W. L. HILL.

CITY OF WASHINGTON,
District of Columbia, ss

Subscribed and sworn to before me this the 16th day of February, A. D. 1888.

ROBERT R. SHELLABARGER,
Notary Public, D. C.

[Affidavit to accompany Nos. 5602-66.]

I hereby certify that certain samples of lard, numbered two to six inclusive, and marked "From C. H. S. Mixer, Chicago, 2-8-88," were prime steam lard, and of the quality known as standard lard by the Board of Trade of the city of Chicago. Said samples were from five different lots of lard, and were made by as many different packers, and the samples fairly represented the different lots from which they were taken. The said five samples of lard were drawn on the eighteenth day of February last, and were on the same day shipped per express and were addressed to "Prof. Sharpless, Riggs House, Washington, D. C."

C. H. S. MIXER,
Chief Inspector of Provisions, Chicago Board of Trade.

Personally appeared C. H. S. Mixer, signer of the foregoing statement, who made solemn oath to the truth of the same, this fifth day of March, 1888.

R. S. WORTHINGTON,
Notary Public.

[Affidavit to accompany No. 5610.]

STATE OF ILLINOIS,
  Cook County, ss:

Geo. H. Webster, being first duly sworn, on oath deposes and says, that he is a member of the firm of Armour & Co., doing business in the city of Chicago, Cook County, Illinois, and that he makes this affidavit on behalf of himself and his copartners in said firm; that he has seen and knows the five pound tin of lard which is herewith submitted to Dr. H. W. Wiley, chemist of the Department of Agriculture of the United States, for analysis, and that the same was manufactured by the said firm of Armour & Co., in accordance with their regular formula for the manufacture of refined lard for foreign trade.

GEO. H. WEBSTER.

Sworn and subscribed to before me this 11th day of February, 1888.

CHARLES F. LANGDON,
Notary Public in and for Cook Co., Ills.

TABLE No. 17.—*Pure lards.*

No.	Specific gravity at 35°	Melting point °C.	Melting point of fat acids. °C.	Crystal izing temp. in three of fat acids. °C.	Color reactions with—		Refractive index at 25°	Rise of temperature with H₂SO₄. °C.	Saponification equivalent.	Reaction with silver nitrate.			Water.	Microscopic indications.	Iodine absorbed.
					H₂SO₄	HNO₃				Bechi.	Milliau.				
															P. ct.
5550	.9354	39.6	42.0	39.6			1.4617	42.1	275.82	Light red	Trace of red action.		Lard stearine	58.35
5565	.9355	39.6	41.6	39.8			1.4620	39.8		...do...	...do...	.615	...do...	59.69
5586	.9357	41.6	43.0	40.4			1.4605	39.7	272.64	Trace of color	...do...	.165	...do...	58.70
5591	.9358	41.2	43.0	41.2	Trace of pink	Light color	1.4625	41.1		...do...	...do...	.090	Lard stearine	58.43
5592	.9363	43.6	45.4	41.3	Faint pink	Trace	1.4622	41.4		...do...	...do...		Lard, possibly tallow stearine	56.93
5593	.9365	45.1	46.6	41.5	Trace of color	...do...	1.4626	40.9		Faint brown	...do...		Lard stearine	60.94
5604	.9371	41.5		40.0	...do...	...do...	1.4633			...do...	...do...		Lard, possibly tallow stearine	61.28
5605	.9367	39.8	41.4	38.9	Faint brown pink	Light pink		41.9		Light brown	Trace of reduction		Lard stearine	60.02
5634	.9351	40.4	43.6	40.0	Light brown pink	Faint color	1.3640			Light red brown	No reduction			60.59
5655	.9358	41.9	42.8	40.4	Light yellow		1.4628	37.1	281.01	Faint red	Trace of reduction	.625	...do...	51.04
5656	.9342	34.7	40.2	42.7	Marked red brown	Slight coloration	1.4621	42.9	280.85	Light brown	...do...	.695	...do...	60.09
5657	.9657	42.9	42.2	39.7	Trace of yellow	...do...		40.8	281.14	No color	...do...	.630	...do...	55.85
5658	.9656	39.0	41.4	36.6	Trace	No color		42.6	282.05	...do...	...do...	.925	...do...	61.45
5672	.9556	35.5	40.6	36.4			1.4623	46.5	283.16	Faint color	...do...		...do...	55.05
5673	.9555	41.0	42.2	42.5			1.4646	36.6	274.12	Marked red	...do...		...do...	55.34
5674	.9552	42.3	41.0	41.3			1.4617	43.5	278.35	Slight color	...do...		...do...	62.55
5676	.9584	37.1	36.9	32.1			1.4659	42.5		No color	...do...			77.78
5679	.9567	36.0	48.4	38.0			1.4648	41.5	280.93	Slight color	Jelly	.617		63.98
Means	.9652	39.7	42.3	39.6			1.4630							62.48

NOTES ON TABLE NO. 17.*

This table includes the analyses of nineteen samples of lard, which both in pedigree and properties appear to be pure hog grease, taken from those parts of the animal usually devoted to lard-making.

Under the head of miscellaneous lards, there are other samples which appear to be pure lard, but the evidence was not in all cases sufficiently conclusive to warrant their incorporation in this table.

Number and description of samples in Table No. 17.

Number.
5550. "Pure Natural Lard," brand of Jacob C. Shafer & Co.; purchased in Washington. Affidavit of Walter L. Hill.
5565. "Best Refined Lard," brand of G. Cassard & Son, Baltimore, Md.; purchased in Savannah, Ga. Affidavit of Isaac G. Haas.
5566. "Leaf Lard," brand of Rohe & Bro., New York; purchased in Savannah, Ga. Affidavit of Isaac G. Haas.
5591. "Pure Leaf Lard," brand of John P. Squire & Co., Boston, Mass.; from manufacturer.
5592. Same as above.
5593. Same as above.
5600. "Pure Unadulterated Lard," brand of F. Whittaker & Sons, Saint Louis, Mo. Affidavit of R. A. Hamilton.
5601. "Pure Unadulterated Honest Refined Lard," brand and affidavit same as above.
5606. "Best Refined Lard," brand of G. Cassard & Sons; purchased in Washington. Affidavit of Walter L. Hill.
5602. "Pure Country Lard," rendered by L. Entriken, West Chester, Pa.
5655. "Pure Leaf Lard," from Deerfoot Farm Company, Boston, Mass. Affidavit of Frank W. Bennett.
5656. "Choice Leaf Lard," brand of Charles H. North & Co., Boston, Mass.; purchased of manufacturer. Affidavit of Frank W. Bennett.
5656. No brand, leaf lard, of Niles Brothers, Boston, Mass. Affidavit of Frank W. Bennett.
5657. "Pure Leaf Lard," brand of Sperry & Barnes, New Haven, Conn.; purchased in Boston, Mass.
5672. Lard from head of hog, rendered in United States Department of Agriculture.
5673. Intestine lard rendered in laboratory.
5674. Leaf lard, rendered in United States Department of Agriculture.
5676. Lard from pigs' feet, from David Wesson, Chicago, Ill.
5679. "Pure Natural Lard," brand of Jacob C. Shafer & Co.; purchased from manufacturer.

In specific gravity 5655 resembles lard stearine. Pigs' feet lard, 5676, should be considered apart, since no lard of commerce is ever made exclusively of pigs' feet. It represents the other extreme of specific gravity.

The highest melting point is shown by 5593 and the lowest by 5676. The highest color with acids was shown by 5656 and with silver nitrate by 5673.

Judged by density alone, 5655 and 5676 would have pronounced adulterated, the former with stearine, the latter with cotton oil. The latter sample would also be looked on as suspected by reason of its high re-

---

* In serial order from page 120, part first.

fractive index. Both 5676 and 5672 show iodine numbers which would lead the analyst to look for a high percentage of cotton oil.

Leaving out these samples made from special parts of the animal, the mean iodine number for the other samples would be materially reduced.

It appears that the true number for lards of commerce would be about 60.

In addition to its low specific gravity No. 5655 is abnormal, both in the slight rise of temperature it gives with sulphuric acid and its low iodine number. In all three properties, viz, specific gravity, rise of temperature with sulphuric acid, and iodine number, it indicates the presence of lard stearine, or that it is made from some special part of the fat.

Prime steam lard. Table No. 18.

No.
5629. Prime steam lard purchased in Saint Louis, Mo. Affidavit of R. A. Hamilton.
5639. Prime lard, steam-rendered, from D. E. Fox, taken by Chicago inspector. Affidavit of Carl Dreier.
5640. Prime steam lard from D. E. Fox, taken by Chicago inspector. Affidavit of Carl Dreier.
5641. Prime steam lard from D. E. Fox, taken by Chicago inspector. Affidavit of Carl Dreier.
5642. Prime steam lard from D. E. Fox, taken by Chicago inspector. Affidavit of Carl Dreier.
5650. Prime steam lard from John P. Squire & Co., Boston, Mass.
5662. Prime steam lard taken by Chicago inspector. Affidavit of C. H. S. Mixer.
5663. Same as above.
5664. Same as above.
5665. Same as above.
5666. Same as above.

TABLE No. 18.—*Steam lards.*

No.	Specific gravity at 15° C.	Melting point. C°	Melting point of fat acids. C°	Crystallizing temperature of fat acids. C°	Color reactions with— H₂SO₄	Color reactions with— HNO₃	Refractive index at 25° C.	Rise of temperature with H₂SO₄. C°	Saponification equivalent.	Reaction with silver nitrate— Bechi	Reaction with silver nitrate— Milliau	Water	Microscopic indication	Iodine absorbed
												Per ct.		*Per ct.*
5629	.9652	38.4	41.8	39.53	Slight color	Light yellow	1.4616	32.7	275.86	Faint brown	Trace of reduction	.080	Lard stearine	63.84
5629	.9651	39.0	41.8	38.93	Very light color	Light red	1.4612	40.0	286.84	Light brown	do	.130	do	62.11
5640	.9654	39.5	42.6	39.65	Light red	Marked pink	1.4617	39.6	276.11	do	Jelly	.110	do	62.33
5641	.9651	42.9	42.0	35.40	Faint color	Bright pink	1.4638	38.8	288.87	Faint brown	Trace of reduction	.140	do	66.47
5642	.9609	32.3	42.2	39.00	Trace	Marked pink	1.4622	42.1	282.42	Slight discoloration	Jelly	.140	do	62.25
5650	.9616	37.7	41.4	38.49	Light red-brown	Pale yellow-brown	1.4646	41.1	277.20	Slight color	Trace of reduction	.040	do	62.22
5662	.9646	38.3	42.0	39.30	Trace	Faint pink	1.4675	41.6	286.87	do	do	.055	do	61.95
5663	.9663	29.8	42.2	38.03	do	do	1.4641	41.3	280.65	do		.235	Lard stearine	63.82
5664	.9659	31.5	42.0		do	Trace	1.4631	29.3		do	Trace of reduction			60.34
5665	.9667	38.9	42.4	29.10	do	do	1.4636	41.5	288.61	do	do	.130	do	63.12
5666	.9656	37.0	42.1	38.60		do	1.4623	39.9	284.45			.160		62.86

Notes on Table 18.

As can be seen by the description of the samples, Table No. 18 contains analyses of fairly good specimens of the prime steam lard of the Chicago market.

The specific gravities of the samples are very near together, differing in any case at most only .0014 from the mean.

The variations in the melting point are more marked, and in Nos. 5663 and 5664 we notice results which are quite anomalous. In No. 5663 the melting point and crystallizing point of the fat acids are comparable with the mean results, which leads to the suspicion of some inadvertent error in determining the melting point of the glycerides. The mean refractive index is slightly higher than that for lards made in other ways. The iodine number is also higher than for pure lards of different origin, especially with the exceptions noted in table No. 18.

When the rise of temperature with sulphuric acid, however, is considered, a lower number is obtained than in No. 17. The numbers for single samples show a close agreement with the exception of 5629, one of the two samples in the table not obtained in Chicago.

As a general observation it may be stated that the steam lards of commerce have a more constant composition than pure lards made in other ways and from more restricted portions of the animal.

Steam lards have a distinctively strong odor which distinguishes them from lards rendered in open kettles at low temperatures and from selected portions of the fat.

Cottonseed oil.

No.
5553. Cotton oil, from F. Whittaker & Sons, Saint Louis, Mo.
5554. Yellow cotton oil, from Naphey & Co., Philadelphia, Pa.
5555. White or refined cotton oil, same source as above.
5615. Summer yellow, received from D. E. Fox.
5616. Summer white, received from D. E. Fox.
5618. From Z. D. Gilman, Washington, D. C., marked Olive Oil Sublime.
5619. Cotton oil, same source as above.
5625. Cotton oil, purchased in Boston, Mass. Affidavit of Walter L. Hill.
5628. Cotton oil, purchased in Saint Louis, Mo. Affidavit of D. H. Kennett.
5645. Prime cotton oil, from N. K. Fairbank & Co., Chicago, Ill. Affidavit of Carl Dreier.
5647. Light yellow cotton oil, Maginnis Oil Works, New Orleans, La. Affidavit of Carson Mudge.
5648. Light yellow cotton oil, purchased from Union Oil Company, New Orleans, La. Affidavit of Carson Mudge.
5649. Light yellow cotton oil, purchased from Delta Oil Works, New Orleans, La. Affidavit of Carson Mudge.
5661. Cotton oil, taken from car by C. H. S. Mixer, in Chicago, Ill. Affidavit of C. H. S. Mixer.
5683. Summer yellow cotton oil, from Southern Cotton Oil Trust.
5684. Summer white cotton oil, Southern Cotton Oil Trust.
5685. Winter yellow cotton oil, Southern Cotton Oil Trust.
5686. Winter white cotton oil, Southern Cotton Oil Trust.

TABLE No. 19.—*Cotton oils.*

No.	Specific gravity at 35°	Melting point of fat acids, C°	Crystallizing temperature fat acids, C°	Color reactions with—		Refractive index at 25°	Rise of temperature with H₂SO₄, C°	Saponification equivalent.	Reaction with silver nitrate.		Water.	Iodine absorption method.
				H₂SO₄	HNO₃				Broth.	Milium.		
											Per cent.	Per cent.
5553	.9149	37.2	32.8			1.4671	84.6		Blackened	Marked reduction	.125	112.99
5554	.9139	38.6	35.1	Green-brown		1.4669	82.1		Very black	do	.655	110.95
5555	.9134	39.4	34.5	Marked brown-red	Marked red-yellow	1.4677	85.8		Black-brown green tint	Slight reduction		109.28
5616	.9149	39.6	35.1	Green-yellow-brown	do	1.4677	86.3		do	Heavy reduction		108.72
5618	.9143	34.6	32.4	Brown-red	Yellow-red	1.4677	88.4		do	No reduction		105.59
5619	.9149	36.6	32.1	Very dark red-brown	Dark yellow-red	1.4682	90.2		Very black	do		111.46
5625	.9152	36.6	32.3	Dark red-brown	Deep red	1.4683	88.8		Very black-brown lime tint	Trace		113.39
5626	.9154	37.4	30.9	Marked red-brown	Red-brown	1.4681	84.3					105.31
5645	.9149	42.6		Marked brown	Light red	1.4685	98.1	288.25	Very black-brown	Jelly	.059	116.97
5647	.9136	41.4	35.6	Red-brown	Red-brown	1.4670	88.4	278.75	Black	Slight reduction	.070	105.68
5648	.9137		35.2	do	do	1.4670	84.9	286.80	do	do	.035	105.42
5649	.9131	42.5	35.2	Light yellow	do	1.4681	84.9	282.05	do	Jelly	.059	109.94
5661	.9141	44.4	34.4		Marked yellow	1.4705	87.3	287.03	Dark brown	No reduction	.020	108.02
6563	.9131	33.8	34.6			1.4649	85.4		Deep red-brown	Heavy reduction		108.59
5684	.9152	36.8	30.5			1.4675	84.4	286.10	do	Slight reduction		106.19
5685	.9154		31.8			1.4680	85.5	253.59	do	do		114.49
5686	.9128	35.8								No reduction		111.06
Means.	.9142	38.8	34.5			1.4658	86.7	283.80			.061	109.04

LARD AND LARD ADULTERATIONS. 485

NOTES ON TABLE No. 19.

The cotton oils examined are believed to represent very accurately the oils used in the adulteration of lards. The samples were mostly taken from large reservoirs and hence better represent a mean value than if derived from small quantities of the material.

The specific gravity of the samples is remarkably uniform, the greatest variation from the mean being $+.0011$.

The high melting and crystallizing points of the fat acids are remarkable characteristics when the low temperature at which cotton oil is a solid is taken into consideration. The figures show how independent these acids are of the glyceride in many of their physical properties.

The high refractive index of cotton oil has already been noted. In No. 5661 this index is far above the mean, while in No. 5649 it falls considerably below. With these exceptions there is a fair agreement among the indices of the remaining samples.

The great rise of temperature shown by cotton oil in contact with sulphuric acid is fully illustrated by the numbers in the table. These numbers are fairly concordant. The greatest departures from the mean are $-6.3°$ and $+3.5°$.

By the silver nitrate test the original samples were easily recognized as cotton oil, while with the same test applied to the free acids, the results, as already indicated, were not so decisive. The probable reason for this has already been mentioned.

In the samples marked "jelly" in all the tables the silver test would not work on account of a gelatinous precipitate, due doubtless to the formation of a salt in the samples, arising from the union of an organic acid with the silver. This organic acid was separated, but not in sufficient quantity to determine its properties. The high iodine number is another characteristic to be noted. Nos. 5649 and 5645 show the greatest departures from the mean.

Stearines.

No.
5612. Prime oleo-stearine, made and used by Armour & Co., Chicago, Ill. Affidavit of George H. Webster.
5613. Prime lard stearine, made and used by Armour & Co., Chicago, Ill. Affidavit of George H. Webster.
5626. Oleo-stearine from John Rearden & Sons, Boston, Mass. Affidavit of Walter L. Hill.
5630. Yellow cottonseed oil stearine, brand of N. K. Fairbank & Co., Chicago, Ill., from E. Richards.
5631. Cotton-seed stearine obtained by Z. D. Gilman, from E. Richards.
5643. Prime lard stearine, from N. K. Fairbank & Co., Chicago, Ill. Affidavit of Carl Dreier.
5644. Oleo-stearine, from N. K. Fairbank & Co., Chicago, Ill. Affidavit of Carl Dreier.
5672. Dead-hog stearine, from John P. Squire, Boston, Mass.
5675. Sample from David Wesson, supposed to be cottonseed oil stearine.
5680. Stearine from white cottonseed oil, from Southern Cotton Oil Trust.
5681. Stearine from yellow cottonseed oil, from Southern Cotton Oil Trust.

LARD AND LARD ADULTERATIONS. 487

TABLE No. 20.—*Stearines used in lard adulteration.*

No.	Specific gravity at 25°.	Melting point, C°.	Color reactions with—		Refractive index at 25°.	Rise of temperature with H_2SO_4, C°.	Saponification equivalent.	Reaction with silver nitrate.		Water.	Microscopic indication.	Iodine absorbed.
			H_2SO_4.	HNO_3.				Becchi.	Millian.			
										Per ct.		Per ct.
5612	.9006	51.9	None	Trace of yellow				No color		.940	Tallow stearine	17.38
5613	.9031	44.3	Trace	Trace	*1.4591	24.7		Faint blue	No reduction	.050	Lard stearine	44.24
5626	.9011		Marked pink	Light yellow	1.4579			Faint pink	No reduction	.295	Tallow stearine	26.81
5629	.9175		Dark brown	Quite red				Deep brown-red	Slight reduction, red-brown			89.54
5631	.9112		...do	Red yellow				Very black brown				85.28
5642	.9039	45.8	None	Trace of pink		27.7		Faint color	No reduction	.035	Lard stearine	49.78
5644	.8998	53.8	Trace	Light yellow	*1.4581	20.0		No color	...do	.320	Tallow stearine	16.00
5652	.9091	58.2	Very dark brown	Very dark red				...do	Mere trace	.850		45.25
5675	.9119				1.4682	80.4		Deep red-brown	Light reduction			92.78
5680					1.4664			...do				59.29
5681					1.4659	64.0	263.11	Very black	Heavy reduction, very black			91.90

* See notes.

Notes on Table No. 20.

The number of stearines examined is not large enough to fix a standard of comparison. Those of interest in the study of mixed lard are the prime lard and oleo-stearines. The cotton-oil stearines are not very extensively used in lard adulterations. The dead hog-grease stearines are never used for this purpose.

The specific gravity of the oleo-stearines is slightly below that of the prime lard stearine samples, and both are less than the specific gravity of pure lard. The melting point of both stearines is above that of lard. In the column containing the refractive indices those marked with an asterisk are of samples of different origin from the remainder of the table, but they are believed to be good representative samples. The mean index, however, should be determined from a much larger number of samples.* It will doubtless be found to be somewhat lower than the index of pure lards taken at the same temperature.

The rise of temperature on mixing with sulphuric acid is much less in the case of the stearines than with pure lard, showing that this phenomenon is chiefly characteristic of oleine. The stearines of cotton oil, however, show an increase of temperature comparable with that of the original oil, and lard stearine a much greater increase than oleo.

The low iodine equivalent of oleo-stearine has already been noticed and is strikingly shown by the data in the table. The cotton-oil stearines show a marked decrease from the numbers obtained for the oil itself.

Armour's lards.

No.
5552. Pure refined family lard, Washington, D. C. Affidavit of W. L. Hill.
5557. Kettle refined lard, Mobile, Ala. Affidavit of F. H. McLarney.
5559. Pure refined family lard, Macon, Ga. Affidavit of T. Skelton Jones.
5561. Choice refined family lard, Macon, Ga. Affidavit of T. Skelton Jones.
5562. Choice refined family lard, Kansas City, Mo. Affidavit of T. Skelton Jones.
5564. Pure refined family lard, Savannah, Ga. Affidavit of Isaac G. Haas.
5572. Choice family lard, Saint Louis, Mo.
5584. Choice family lard, Kansas City, Mo. Affidavit of E. K. Converse.
5584. Pure refined family lard, New Orleans, La. Affidavit of E. K. Converse.
5595. Pure refined family lard, Philadelphia, Pa. Affidavit of W. L. Hill.
5610. Pure refined family lard. Affidavit of George H. Webster.
5611. Pure refined family lard. Affidavit of George H. Webster.
5653. Superior compound lard, Boston, Mass. Affidavit of Frank W. Bennett.

---

* The analytical data show that oleine has a higher refractive index than stearine or palmitine.

LARD AND LARD ADULTERATIONS.

TABLE No. 21.—*Armour's lards*.

No.	Specific grav. ity at 26°.	Melting point, C°.	Melting point of fat acids, C°.	Crystal- lizing tempera- ture of fat acids, C°.	Color reactions with—		Refrac tive index, at 25°.	Rise of tempera- ature with H₂SO₄ C°.	Saponi- fication equiva- lent.	Reactions with silver nitrate.		Water.	Microscopic in- dications.	Iodine ab- sorbed.
					H₂SO₄	HNO₃				Becchi.	Millian.			
												Per ct.		Per ct. 71.19
5552	.9064	39.7	42.0	38.5			1.4668	42.1	273.96	Almost black		.110	Lard and tallow stearine.	61.18
5557	.9061	43.3	42.0	41.1			1.4626	44.3	273.32	Slight color	Slight reduction	.025	Lard stearine	62.68
5559	.9057	39.5	42.6	39.9			1.4629	42.5		Quite brown	Jelly	.050	...do	61.67
5561	......	41.9	42.4	......			1.4652	38.9	276.37	Very black		.360	Tallow stearine	61.67
5562	.9057	41.7		40.1			1.4617	42.1	273.44	Brown	Marked reduc- tion.			
5564	.9063	38.9	42.0	39.3			1.4616	42.5	275.71	Quite brown	Marked reduc- tion.	.120	Lard stearine	61.30
5573	.9045	42.5	41.8		Dark brown	Dirty brown	1.4646	40.9	272.84	Red-brown		.055	Tallow stearine	54.11
5581	.9063	40.7	42.4	39.3	Light red-brown	Red-brown	1.4622	44.7	276.66	Black-green tint	Slight reduction	.029	Lard and tallow stearine.	63.97
5584	.9066	39.5	42.6	39.4	Slight color	Faint red	1.4628	56.1	279.42	Brown-red	do	.105	do	66.73
5595	.9065	38.9	42.4		Dirty brown	Yellow red	1.4618	41.0		Black brown- green tint.			do	67.32
5610	.9064	41.9		39.8	Marked brown	do	1.4626	46.0		do	Trace of reduc- tion.			65.41
5611	.9065	40.0		39.5	Light brown	Light yel- low red.	1.4629	44.7		do				62.00
5653	.9055	39.7	43.4	40.5	Quite brown	Pink red	1.4638	41.5		Dark brown				65.46
Means	.9060	40.6	42.8	39.8			1.4634	46.5	275.23			.098		63.38

Notes on Table 21.

The samples analyzed show that the compound lards manufactured by Armour & Co., of Chicago, have nearly a uniform constitution. The specific gravities of the various samples differ only slightly from the mean. The maximum difference is in No. 5572, viz, — .0015, and the maximum + difference in No. 5611, viz, .0008. The melting points are also remarkably constant, the two maximum variations of a positive and negative sign being + 2.7°, in No. 5557 and — 1.7° in Nos. 5564 and 5595. The same agreement is also noticed in the melting and crystallizing points of the fat acids. The refractive index in most cases is only slightly above that of pure lard, showing only a small addition of cotton oil, or a correction of the index thereof by a corresponding addition of a stearine with low index.

The rise of temperature with sulphuric acid shows only notable variations in Nos. 5561 and 5584.

In the latter of these the iodine number is correspondingly increased, but not so in the former. This is only another of the numerous illustrations of the analytical perplexities encountered in the study of mixed glycerides.

The reactions with silver nitrate reveal very well, in almost every instance, the presence of cotton oil, but in some cases the phenomena of reduction of silver are not sufficiently developed to distinguish the samples from lard containing enough of impurity other than cotton oil to give a color with silver nitrate. The silver test applied to the free acids did not afford satisfactory results; a fact which has already been noted, and its possible explanation given.

In the iodine numbers, those obtained for Nos. 5552, 5584, and 5595 are much above the mean, while in one case, No. 5572, the percentage of iodine absorbed, viz, of 54.11, would indicate the admixture of a larger quantity than usual of oleo-stearine.

Fairbank's lard.

No.
5558. Prime refined family lard, purchased in Mobile, Ala. Affidavit of F. H. McLarney.
5561. Prime refined family lard, purchased in Macon, Ga. Affidavit of T. Skelton Jones.
5563. Choice refined family lard, purchased in Savannah Ga. Affidavit of Isaac G. Haas.
5567. Prime refined family lard, purchased in Dallas, Tex. Affidavit of Thomas F. McEnnis.
5569. Prime refined family lard, purchased in Saint Louis, Mo.
5573. Prime refined family lard, purchased in Saint Louis, Mo.
5574. Prime refined family lard, purchased in Atlanta, Ga.
5576. Prime refined family lard, purchased in New Orleans, La. Affidavit of E. K. Converse.
5586. Prime refined leaf-lard, purchased in Norfolk, Va. Affidavit of W. B. Pearman.

5596. Prime refined family lard, purchased in Philadelphia, Pa.   Affidavit of W. L. Hill.
5634. No brand; original small package "Y."   Affidavit of William T. Wells.
5635. No brand; original large package "Z."   Affidavit of William T. Wells.
5636. "X" prime refined family lard, purchased in New York.   Affidavit of William T. Wells.
5637. "XX" prime refined family lard, purchased in New York.   Affidavit of William T. Wells.
5638. "S" Cuba export refined lard, purchased in New York.   Affidavit of William T. Wells.
5646. Prime refined family lard, from D. E. Fox.   Affidavit of Carl Dreier.
5654. Compound lard purchased in Boston, Mass.   Affidavit of Frank W. Bennett.

492 FOODS AND FOOD ADULTERANTS.

TABLE 22.—N. K. Fairbank's Lards.

No.	Specific gravity at 35°C	Melting point, °C	Melting point of fat acids, °C	Crystallizing temperature of fat acids, °C	Color reactions with—			Refractive index at 35°	Heating temperature with H_2SO_4 (Mod.)	Saponification equiv., about	Reaction with silver nitrate.			Water.	Microscopic indication.	Iodine absorbed.
					H_2SO_4	HNO_3					Becchi.		Miliau.			
														P. ct.		P. ct.
5556	.9113	36.4	40.2	36.7				1.4656	58.1	276.37	Very black		Heavy reduction, black.	.28	Tallow stearine	86.36
5561	.9099	41.9	41.8	37.5				1.4652	58.9	276.37	...do...		Marked reduction.	.26	...do...	83.28
5562	.9164	38.8	40.2	38.5				1.4700	58.2	276.93	Quite black			.24	...do...	92.62
5567	.9168	38.1			Brown	Red brown		1.4692	62.9	274.57	...do...		Slight reduction	.31	...do...	93.31
5569	.9123	34.4		35.7	Light brown	Light red-brown		1.4653	61.5	274.57	Green-black			.25	...do...	92.22
5573	.9169	31.3		36.0				1.4663	54.2		...do...			.13	Lard and tallow stearine.	82.47
5574	.9100	12.5	20.0		Very dark brown	Pronounced red		1.4626	54.3	277.20	...do...		Slight reduction	.05	Tallow stearine	79.47
57..	.9100	13.1	40.0		Decided brown	Dirty brown		1.4639	69.8	276.26	Very black		Marked reduction.	.27	...do...	94.76
5585					...do...	Deep brown red					Green-black			.20	Lard stearine	88.92
5591	.9494	29.7		33.4	Dirty brown	Yellow-red		1.4648	52.0	276.96	Black-brown		Heavy reduction		Tallow stearine	88.17
5594	.9087	29.5	42.4	29.5	Red-brown	Red		1.4631	56.5		Deep black, green tint.				...do...	83.89
5425	.9322	39.9	41.6	37.7	...do...	Light red		1.4632	55.0	281.38	...do...		Marked reduction.	.69	...do...	78.24
5656	.9684	25.9	24.4	31.7	Light red-brown	...do...		1.4696	57.9	281.69	...do...		Heavy reduction		...do...	83.93
5687	.9886	25.3	40.0	27.9	...do...	Marked red		1.4648	59.2	240.50	...do...		Marked reduction.		...do...	87.22
5628	.9675	42.9	32.5	32.5	Red-brown	Red		1.4654	52.9	240.16	Black-brown				...do...	75.57
5640	.967.	40.4	42.2	30.3	Brown	Yellow-brown		1.4629	33.1	240.49	Dark brown, green tint.		Marked reduction.		...do...	76.24
5654	.9068	38.9	45.1	42.9	Marked brown	Red-brown		1.4644	39.6	252.15	Black		Slight reduction	.65	Tallow stearine	85.60
Means	.9305	30.1	40.6	37.4				1.4651	55.9	270.40				.195		84.31

NOTES ON TABLE NO. 22.

This table presents the results of an interesting study of compound lards in which the natural hog grease is reduced to a minimum.

Indeed it appears from the general results in some of the samples that they may not have any lard in them at all, but lard stearine instead. The high specific gravity, low melting point, high refractive index, great rise of temperature with sulphuric acid, and high iodine number, all point to samples containing a maximum quantity of cotton oil or cotton-oil stearine.

The chief variations from the mean of the specific gravity are shown in Nos. 5569 and 5635. In both cases the iodine numbers conform to the indications of the density.

In one case (No. 5573) the melting point is very low, while the highest melting point is (5638) a compound lard made for the Cuban trade, and having, therefore, presumably a large content of oleo-stearine, and in which we might expect the cotton oil to be present as a stearine also.

The refractive indices reveal unmistakably the presence of a body with a higher index than pure lard, and the high temperatures reached with sulphuric acid are a further evidence that this substance is cotton oil. The iodine numbers furnish the rest of the evidence, showing the high percentage of this substance present in the mixture.

The results shown in this table are much more satisfactory than those recorded in the preceding one. It is quite evident that some samples of Armour's lards might pass as pure hog grease, when the microscope fails to reveal crystals of beef fat, while none of the Fairbank samples could be thus mistaken.

Miscellaneous Lards.

No.
5551. "Cream Leaf Lard," brand of Charles G. Kriel; purchased in Washington, D. C. Affidavit of Walter L. Hill.
5556. "Choice Family Lard," brand of Kansas City Packing Company, Kansas City, Mo.; purchased in Mobile, Ala. Affidavit of F. H. McLarney.
5560. "Choice Refined Family Lard," brand of Jacob Dold & Co., Kansas City, Mo.; purchased in Macon, Ga. Affidavit of T. Skelton Jones.
5568. "Choice Family Lard," brand of Fowler Brothers, Chicago, Ill.; purchased in Saint Louis, Mo.
5571. "Cream Leaf Lard," brand of Charles G. Kriel, Baltimore, Md.; purchased in Saint Louis, Mo.
5575. "Choice Family Lard," brand of Anglo-American Provision Company, Omaha, Nebr.; purchased in Atlanta, Ga.
5577. "Choice Family Lard," brand of George Fowler, Kansas City, Mo.; purchased in New Orleans, La. Affidavit of E. K. Converse.
5578. "Choice Refined Family Lard," brand of Allcutt Packing Company, Kansas City, Mo.; purchased in New Orleans, La. Affidavit of E. K. Converse.
5579. "Choice Family Lard," brand of Kansas City Packing Company, Kansas City, Mo. Affidavit of E. K. Converse.
5580. "Pure Refined Family Lard," Plankinton Company brand, Milwaukee, Wis.; purchased in New Orleans, La. Affidavit of E. K. Converse.

5585. "Anchor Lard," brand of A. H. Worthman & Co., Philadelphia, Pa.; purchased in Norfolk, Va. Affidavit of C. A. Woodard and W. B. Pearman.
5587. "A No. 1 Refined Lard," brand of Swift & Co., Chicago, Ill.; purchased in Norfolk, Va. Affidavit of C. A. Woodard and W. B. Pearman.
5588. "Pure Refined Lard," brand of Chicago Packing and Provision Company; purchased in Norfolk, Va. Affidavit of C. A. Woodard and W. B. Pearman.
5589. "Kettle-Rendered Leaf Lard," brand of Robe & Bros., New York; purchased in Norfolk, Va. Affidavit of C. A. Woodard and W. B. Pearman.
5590. "Choice Family Lard," purchased in Boston.
5594. Brand: "Pure Family Lard," Halstead & Co., New York. Affidavit of W. L. Hill.
5597. Brand: "(Marca Castellana) Mantica pura Calla Forsyth," New York. Affidavit of W. L. Hill.
5598. Brand: "Choice Refined Family Lard," Allcutt Packing Company, Kansas City, Mo.; purchased in Dallas, Tex. Affidavit of Thomas F. McEnnis.
5599. "Choice Family Lard," brand of Charles F. Tietjen, from Naphy & Co., Philadelphia.
5602. Adulterated Refined Lard, purchased in Saint Louis, Mo. Affidavit of R. A. Hamilton.
5632. Received from D. E. Fox, from Charles F. Tietjen, for Central Lard Company, New York. Affidavit of Charles F. Tietjen.
5633. "Marked B, same as A," from D. E. Fox.
5667. Kettle-rendered, backs and leaf lard, from Plumb & Winter, Bridgeport, Conn.
5668. Kettle-rendered, backs and leaf lard, from F. A. Bartran & Co., Bridgeport, Conn.
5669. Kettle-rendered, intestinal and head lard, from F. A. Bartran & Co., Bridgeport, Conn.

LARD AND LARD ADULTERATIONS 495

TABLE No. 93.—*Lards of miscellaneous origin.*

No.	Specific gravity at 35°.	Melting point, C°.	Melting point of fat acids, C°.	Crystallizing temperature of fat acids, C°.	Color reactions with—			Refractive index at 25°.	Rise of temperature with H₂SO₄ C°.	Saponification equivalent.	Reaction with silver nitrate.		Water.	Microscopic indications.	Iodine absorbed.
					H₂SO₄.	HNO₃.					Berlin.	Millon.			
													Per ct.		Per ct.
5351	.9361	37.7		38.3				1.4529	43.5	271.42	Light red	Slight reduction	.915		66.42
5356	.9357	40.0	43.2	39.9				1.4639		272.43	Marked brown		2.000	Lard stearine	62.09
5362	.9563	42.7	44.0	41.0	Brown	Red-brown		1.4723	50.6	271.66	Slightly brown	No reduction	.070	Tallow stearine	57.18
5368	.9572	43.7	42.2	29.6		Red-brown		1.4615	29.5	273.52	Dark red-brown	...do	.621	...do	66.31
5371	.9591	39.1	39.2	37.0	Very deep brown	Muddy brown		1.4615			Red-brown, doubtful		.153	Lard and tallow stearine	61.54
5375	.9551	44.5			Pink, then brown	Muddy pink		1.4608	37.7		Brown-black, light-green tint		.120	Tallow stearine	61.76
5577	.9956	40.1	42.9	40.2	Faint muddy pink	Marked pink		1.4635	41.9	272.68	Light brown, doubtful		.305	Lard and tallow stearine	76.91
5578	.9651	40.1	42.0	39.6	Pink brown			1.1065	50.1		Light brown		1.565	...do	63.67
5579	.9567	36.0	40.6	37.1	Light red-brown	Light red		1.4641	47.5	270.48	Brown-black		2.110	...do	62.17
5580	.9561	44.1	45.2	40.1	Red-brown	Dirty red-brown		1.4618	43.4	279.94	Brown-black, green tint		.620	Tallow stearine	29.99
5585					Trace	Trace					Light red		12.285	Lard, and trace of tallow stearine	
5587	.9959	42.9			Light brown	Pink, then red-brown					Brown-black, green tint		9.040	Tallow stearine	58.89
5588	.9971	42.2			Dark brown	Brown red	1.4672				...do		.145	...do	55.88
5589					Light pink brown	Pink, then red-brown					...do		1.140	...do	61.92
5590	.9960	40.7	44.4	30.5	Light dirty brown	Light red-brown	1.4652	41.8	272.21	...do	Marked reduction		...do	63.61
5591	.9975	42.4		39.7	Dirty brown	Marked yellow red	1.4629	50.3		...do	...do	1.290	...do	74.98

496　　　　　　　　　　　　　　FOODS AND FOOD ADULTERANTS.

TABLE No. 23.—*Lards of miscellaneous origin*—Continued.

No.	Specific gravity at 35°.	Melting point, C°.	Melting point of fat acids, C°.	Crystallizing temperature of fat acids, C°.	Color reactions with—		Refractive index at 35°.	Rise of temperature with H_2SO_4, C°.	Saponification equivalent.	Reaction with silver nitrate.			Water.	Microscopic indications.	Iodine absorbed.
					H_2SO_4	HNO_3				Béchi.	Millian.				
5597	.9072	41.9	......	40.7	Dirty brown	Light red-yellow	......	61.3	......	Brown-black, green tint.	Slight reduction.	Per ct. .120	Tallow stearine	Per ct. 69.57
5598	.9353	42.3	42.8	40.4	Light red-brown	Dirty light brown.	1.4616	42.8	......	do	do	1.335	do	65.43
5599	.9369	42.8	42.0	39.5	Light brown	do	1.4638	51.1	271.82	do	do	25.661	Lard and tallow stearine.	70.62
5602	......	......	......	......	Trace of color	Decided red	......	......	......	Black	......	......	do	74.02
5632	.9050	42.5	......	40.9	Marked red-brown.	Pink, then red	1.4644	49.4	277.14	Brown-black, green tint.	Slight reduction.	.685	Tallow, and probably lard stearine.	65.29
5643	.9066	42.3	......	37.9	Marked brown	do	1.4630	49.0	276.52	Very brown-black, green tint.	......	6.0	Lard and tallow stearine.	68.33
5667	.9015	42.3	41.6	40.8	Trace	Trace	1.4610	41.7	278.86	Light red	Trace of reduction.	.065	Lard stearine	67.27
5668	.9046	38.9	43.2	39.3	do	do	1.4618	42.3	......	do	do	.110	Lard and trace of tallow stearine.	52.79
5669	.9049	41.1	44.0	40.4	do	do	1.4628	33.6	277.95	do	do	1.20	Lard stearine	61.43
Means	.9087	41.7	42.9	39.6	......	......	1.4631	45.7	274.67	......	......	2.830	......	64.34

NOTES TO TABLES NOS. 23, 24, AND 25.

In Table 23 are included all the lards of a miscellaneous origin which have been examined by us, and which could not be properly classified under any of the preceding heads. Of these quite a number deport themselves with re-agents as pure hog grease, while others are without doubt adulterated. Of those which appear pure I will mention Nos. 5551, 5667, and 5669.

The microscope revealed the presence of beef fat in most of the other samples, while other tests confirmed the presence of adulterants.

In Table 24 are collected the data obtained by the analyses of crude cotton oils and foots or tank settlings.

Nos. 5603 and 5604 were furnished as pure cotton oils, but the analyses showed that they were heavily adulterated. These are the only instances in which crude or refined cotton oils were found to be adulterated. The nature of the adulterant was not determined.

Table No. 25 contains the analyses of miscellaneous oils, and especially of dead-hog grease.

The low specific gravity of the lard oil (No. 5621) appears anomalous, since it should be higher than pure lard. Further investigations will determine the normal density of lard oil at any given temperature. Pea-nut oil (No. 5622) has practically the same specific gravity as cotton oil; while olive and rape seed oils have densities slightly above pure lard. Dead-hog lard differs from the pure variety chiefly in its refractive index and the quantity of free acid it contains.

Crude cotton oils and foots.

No.
5570. Cottonseed foots, obtained from Henry Sayers & Co., Saint Louis, Mo.
5582. Crude cotton oil, purchased in New Orleans, La.
5583. Cotton-oil foots, purchased in New Orleans, La.
5603. Summer white cotton oil, from Francis Whittaker & Sons, Saint Louis, Mo. Affidavit of R. A. Hamilton.
5604. Yellow cotton oil; source and affidavit as above.
5605. Crude cotton oil; source and affidavit same as above.
5611. Crude cotton oil; received from D. E. Fox.
5682. Crude cotton oil; from Southern Cotton Oil Trust.
5687. Crude cotton oil, from Brinkley, Ark., obtained from David Wesson, Chicago, Ill.
5688. Crude cotton oil, from Jackson, Tenn.; obtained from David Wesson, Chicago, Ill.

TABLE No. 24.—*Crude cotton oils and foots.*

No.	Specific gravity at 15°	Melting point, C°.	Melting point of fat acids, C°.	Crystallizing temperature fat acids, C°.	Color reactions with—		Refractive index at 25°	Rise of temperature with H₂SO₄, C°.	Saponification equivalent.	Reaction with silver nitrate.		Water.	Iodine absorbed.
					H₂SO₄.	HNO₃.				Bechi.	Millian.		
												Per cent.	*Per cent.*
2573	....	43.6	43.6	....	Dirty yellow-brown.	Dirty yellow.	1.4613	....	....	Very black.	....	33.115	110.69
2582	.9157	....	43.2	35.6	Deep red-brown.	Red brown.	1.4674	72.2	....	Very red, a black.	....	.120	....
2583	.9154	....	39.9	....	...do	Marked red.	1.4679	77.7	....	Deep red.	....	2.850	108.31
2593	....	....	43.6	....	Deep black.	Very brown black.	....	....	....	Marked red.	No reduction.	.735	72.92
2591	....	....	33.4	34.1	Dark brown.	Dark brown.	1.4618	....	....	Dark brown-black.	Trace of reduction.	....	59.81
2595	....	....	38.4	35.1	Deep brown-red.	Deep brown-red.	1.4657	73.8	....	Deep red, but not black.	....	.100	109.96
2614	.9155	....	40.2	34.9	Deep black-brown.	Marked red-yellow.	1.4679	79.0	281.81	Black-brown, green tint.	....	.125	104.25
2642	.9164	....	....	....	....	....	....	....	....	Very black.	....	....	109.45
2651	.9153	....	43.2	33.6	....	....	1.4665	....	....	Deep red-brown.	....	....	107.76
2658	.9154	....	41.3	34.7	....	....	....	....	....	Dark red-brown.	Slight reduction.	....	109.67
Means	.9156	43.6	41.3	34.7	....	....	1.4655	76.9	281.81	....	....	6.180	102.63

Miscellaneous oils.

No.
5617. Marked "Olive Oil Sublime," from Z. D. Gilman, Washington, D. C.
5620. Rape-seed oil, from Z. D. Gilman.
5621. Lard oil, from Z. D. Gilman.
5622. Pea-nut oil, from Z. D. Gilman.
5623. Labeled "Hughes's Extra Superfine Italian Lucca Olive Oil; purchased from Alfred E. Hughes, Boston, Mass. Affidavit of Walter L. Hill.
5624. Olive oil, purchased from Wm. Underwood & Co., Boston, Mass. Affidavit of Walter L. Hill.
5627. Olive oil from Alden Speare's Sons & Co., Boston, Mass. Affidavit of Walter L. Hill.
5651. "Dead-hog grease," from John P. Squire & Co., Boston, Mass.
5659. "Dead-hog grease," from East Saint Louis Rendering Company, East Saint Louis, Ill. Affidavit of R. A. Hamilton.
5660. Dead-hog grease, with 10 per cent. of oleo-stearine added, from East Saint Louis Rendering Company, East Saint Louis, Ill. Affidavit of R. A. Hamilton.
5670. Dead-hog grease from East Saint Louis.
5671. Dead-animal grease from East Saint Louis.
5689. Prime lard oil from David Wesson, Fairbank & Co., Chicago, Ill.

500 FOODS AND FOOD ADULTERANTS.

TABLE No. 25.—*Miscellaneous oils.*

No.	Specific gravity at 15°	Melting of fat acids, C°	Crystalliz- ing temper- ature, fat acids C°	Color reactions with—			Refractive index at 25°	Rise of tem- perature with H_2S O_4 C°	Reaction with silver nitrate.			Water.	Iodine ab- sorbed.
				H_2SO_4	HNO_3				Bechi.	Millian.			
												Per cent.	*Per ct.*
5617	.9056	26.6	19.55	Brown black	Yellow-green		1.4612	59.9	Dirty red-yellow	No reduction			85.37
5620	.9039	21.6		...do	Light red		1.4672	68.0	Light dirty brown	...do		.025	98.72
5621	.8949	42.6		Deep brown	Deep dirty-red		1.4632	54.2	Light brown	...do		1.365	66.36
5622	.9124	32.6	30.87				1.4658	60.1				.125	96.61
5623	.9076			Green, changing to brown.	Light green		1.4643		Light green-brown				83.38
5624	.9060	24.7	17.27	...do	...do		1.4638	45.0	Light green-yellow	No reduction			81.90
5627	.9039	27.9		Deep black-brown	Light yellow		1.4639	51.9	Light red-brown	Mere trace		.100	82.76
5631	.9061	31.4	28.90	Black-brown	Dirty brown		1.4615	44.0	Light red	...do		.225	61.70
5659	.9036	42.9	37.10	Light pink-brown	Dealed pink		1.4618	42.4	No color	Jelly		.155	60.52
5660			39.60	...do	Marked pink				...do			.045	56.13
5670	.9058	40.4	28.30				1.4664	43.7	...do	Mere trace		.245	63.49
5671	.8957	42.6	40.45				1.4585	37.5	Faint yellow	No reduction		3.789	53.42
5680	.9151						1.4650		No color				85.63
Means							1.4638	51.2				.676	73.23

ABSTRACTS OF METHODS USED IN LARD ANALYSIS, WITH RESULTS THEREOF, RECENTLY USED BY ANALYSTS.

Dr. Shippen Wallace[*] has made a study of the adulteration of lard. His conclusions are as follows:

In all samples of suspected lard, if one will follow the method here given, he can not fail to meet with correct and proper results.

(1) Hübl's method, which will indicate either adulteration with tallow alone or cotton-seed oil alone, or indicate pure lard.

(2) Use Bechi's test, as described, which will prove the presence or absence of cottonseed oil.

(3) Use the sulphuric-acid test as a further confirmation.

By these last two, if Hübl's method should yield a figure which would classify the suspected lard as pure, one can readily confirm or disprove it, while if Hübl's should indicate cottonseed oil, they would make the proof complete. Lard stearine yields figures, by Hübl's method, within the range of pure lard, and while some manufacturers make use of this article in the manufacture of summer lard, yet it is not an adulteration in the same sense that cottonseed oil and tallow are. I have not mentioned other claimed adulterants of lard, as they are easy of detection; water we sometimes find, one sample I examined containing 11.80 per cent. When this is found it is either caused by carelessness in the manufacture, or is intentional, as it can readily be guarded against.

The percentages of iodine absorbed by sixteen samples of pure and adulterated lard as found by Dr. Wallace are given in the following table:

No.	Pure.	Adulterated.	No.	Pure.	Adulterated.
	Per cent.	Per cent.		Per cent.	Per cent.
1	57.9	58.0	9	61.0	70.2
2	59.6	54.4	10	59.4	67.0
3	59.2	58.1	11	59.5	69.4
4	60.3	60.4	12	59.0	65.1
5	57.7	59.5	13	58.9	68.8
6	61.7	61.8	14	59.1	70.7
7	62.2	60.0	15	60.0	72.0
8	57.4	68.6	16	59.4	66.8

NITRATE OF SILVER REACTION WITH COTTON OIL.

Bizio[†] criticises the report of the Italian commission which recommended Bechi's test for detecting cotton oil.

According to Bizio pure olive oil sometimes produces reduction of silver even when the re-agent is slightly acidified with nitric acid. On the other hand, some samples of cotton oil fail to produce the reduction. Bizio did not take the same care to identify his samples that was used by the commission, and his criticism will not impair the value of the large experience which has shown the practical reliability of the silver test in the detection of cotton oil.

---

[*] Report of the Dairy Commissioner of the State of New Jersey for 1887, p. 16 et seq.
[†] Chem. Central-Blatt, June 23, 1888, p. 873.

CHLORIDE OF GOLD AS A TEST FOR COTTON OIL.

Hirschsohn[*] has recommended the use of auron= chloride for the detection of cotton oil. The reagent is used as follows:

Dissolve one gram of gold chloride in 200cc of chloroform. To 3 to 5cc of the oil add 6 to 10 drops of the re-agent and heat for twenty minutes. Cotton oil will give a beautiful red color.

David Wesson[†] finds that free fatty acids interfere with the delicacy of the reaction, and also rancid lard.

Free acids and rancid lard, on the other hand, do not affect the process of Brullé.

Moerk[‡] has also reported results of this test.

I have tried the reaction of Hirschsohn, and found the purple color produced quite characteristic; but even pure lards give a trace of color, which must not be confounded with the deep coloration produced by cotton oil.

BRULLÉ'S METHOD OF DETECTING ADULTERATIONS IN OLIVE OIL.[§]

The United States consul (Mr. Mason) at Marseilles writes as follows:

Southern France has of late years suffered seriously from the adulteration, or rather the artificial fabrication, of her two principal agricultural products, wine and olive oil. During the recent season of scanty vintages there has grown up in this district an immense manufacture of "piquettes" or raisin wines, which are made by soaking in water, until fermentation takes place, the cheap dried grapes which are imported in such quantities from the Grecian Archipelago and Turkey. These substitutes have so far replaced the real but more costly French wines that now—since the replanted vineyards begin to yield more abundantly—the genuine ordinary wines command only prices which hardly repay the cost of culture. The consumption of vinous beverages among the laboring classes has not diminished, but the cheaper substitute has crowded out the real article, and in behalf of the agricultural class it is proposed to remedy this unnatural difficulty by putting a heavy import duty upon dried grapes from the Levant.

With olive oil the case is similar, but even worse. Only a small portion of France is adapted to olive culture, the entire available district being a strip of dry country less than 20 miles wide along the Mediterranean coast. The tree is of slow growth, and is, moreover, beset by numerous insects and diseases, which, in addition to unfavorable phases of weather, render the yearly olive crop more or less uncertain. Any serious reduction in the annual consumption of olive oil is sufficient to

---

[*] Pharmaceutishe Zeitschrift für Russland, 1888, p. 724, and American Journal of Pharmacy, January, 1889, p. 23.
[†] Letter February 16, 1889.
[‡] American Journal of Pharmacy, February, 1889, p. 65.
[§] The Grocer and Oil Trade Review, February 2, 1889.

reduce its market value below the point of profitable culture. This has been done by the now nearly universal practice of adulterating or diluting the olive oils of Nice and Provence with various seed oils, viz, sesame, peanut, poppy-seed, camomile, and especially cottonseed, which last, by reason of its cheapness, palatable flavor, and difficulty of detection, has of recent years nearly supplanted all the others as an adulterating material. The rank, low-priced olive oils from southern Italy (Bari), Algeria, and Tunis have been brought here in vast quantities, diluted with cotton or sesame, and been consumed and exported wholesale in place of the fine, delicate, high-grade oils of the Var and Bouches-du-Rhône, which have thus been nearly elbowed out of the market. This has so reduced the value of olive oil in southern France that the Government has set itself seriously to the task of providing a remedy. The first step was to discover some method of detecting such adulterations which should be not only exact in its results but sufficiently simple to be practicable for farmers, dealers, and ordinary consumers. It was stated in a report which was made from this consulate in February, 1888, that no such process was then known. As late as May 17 last a meeting of the Scientific and Industrial Society of Marseilles was addressed by Mr. Ernest Millian, an accomplished analytical chemist, who reviewed elaborately all of the known processes and admitted that none of them were sufficiently delicate and exact to detect an adulteration of less than 10 per cent. The "Cailletet" process, which consists in treating the oil with a mixture of sulphuric and nitric acids, had been hitherto generally employed, but this was declared by Mr. Millian untrustworthy unless the degree of adulteration exceeded 20 per cent.

The "Bechi" process, now used by the Italian Government, will detect an admixture of 15 per cent. of cottonseed oil, provided the sample analyzed contains no glycerine, formic acid, or free fatty acids, any one of which, even in minute quantity, is sufficient to mask the chemical reaction upon which the process of Signor Bechi depends. Mr. Millian then described a new method, invented by himself, which consists in treating with heat the saponified products of the oil in alcoholic solution with nitrate of silver. This, however, is a process for the laboratory of the accomplished chemist, and is not adapted to general use. The same is true of the "Levallois" process, which has been used by experts in cases of real importance with more or less questionable results, the analysis in one notable instance having given the same result from a sample of pure olive oil, and another which was known to contain 20 per cent. of cotton-seed.

Finally, as it would seem, the long sought for process has been discovered by Mr. Brullé, chemist of the Agronomic Station at Nice. His discovery was announced to the Academy of Sciences in April last, and has been since subjected to an elaborate series of tests and experiments by a commission specially appointed for the purpose by the Agricultural Society of the Alpes Maritimes. Mr. Brullé began upon the known principle that vegetable oils, when oxidized by the application of certain acids, assume different shades of color. He then hit upon the use of albumen to fix and accentuate these delicate gradations of tint. The report of the commission has recently been published, and gives the process of Mr. Brullé such complete and unqualified indorsement, both for its simplicity and the exactness of its results, that the subject assumes a practical importance not only to the countries which produce olive oils, but to those which, like our own, import them as costly luxuries for general consumption. In its series of experiments at Nice the commission first applied the process of Mr. Brullé to six classes of samples, viz, first, to pure olive oil, then to the same oil with an added admixture of 5, 10, 20, and 50 per cent., respectively, of cottonseed oil, and finally to the pure cottonseed oil itself. When the result had been established by repeated experiments with each grade of samples a fac-simile of the tint produced by each successive degree of adulteration was prepared by dissolving certain pigments in stated quantities of water. Thus the process and a standard system of proofs were put within reach of any person having a good eye for color and a slight familiarity with chemical manipulations.

THE NEW PROCESS.

The process of Mr. Brullé is as follows: Put into a test-tube $1\frac{1}{2}$ grains* of pure albumen (this should be gently heated in the flame of an alcohol lamp to expel any remaining moisture in the albumen which might otherwise modify the exactness of the result), then add 3 cubic centimeters of nitric acid and 10 cubic centimeters of the oil to be tested (the quantity of each ingredient used is, of course, immaterial, provided the above relative proportions are maintained; a test-tube graduated metrically is the most convenient for the purpose); the mouth of the tube is then closed with a cork to prevent the boiling over of the liquid during ebullition, but pierced with a small orifice to permit the escape of vapor, which would otherwise explode the tube. The materials are mixed by shaking, but the nitric acid quickly settles to the bottom. Now warm gently in the lamp the part of the tube containing the oil, then apply the flame to the underlying stratum of acid. A fierce ebullition soon ensues, and when this is at its height plunge the tube into ice water sufficiently cold to chill the contents to 4° C, or its equivalent 40° F. During the cooling process there is developed an oleaginous precipitate, ranging in color from pale yellow to reddish brown, according to the proportion of cotton oil contained in the tested sample. The experiment requires only the simple apparatus above mentioned, and occupies only four or five minutes.

The findings of the commission at Nice are tabulated in its official

---

* 50 mg.

report as follows, the standard tint in each grade being produced by dissolving the stated number of units of each pigment named in 100 units of water. For this purpose ordinary dry-cake water colors are most convenient:

(1) Pure olive oil yields a precipitate tinged like 5 units of Naples yellow dissolved in 100 units of water.

(2) Olive oil containing 5 per cent. of cotton oil yields the tint of 5 units Naples yellow and 5 units of dark chrome yellow in 100 units of water.

(3) Olive oil containing 10 per cent. cotton seed yields a tint equal to 20 units Naples yellow, $6\frac{1}{2}$ units chrome yellow, and 1 unit Chinese vermilion in 100 units of water.

(4) Olive oil containing 20 per cent. cotton seed yields a tint equal to $6\frac{1}{2}$ units Naples yellow, 6 units chrome yellow, and $1\frac{1}{2}$ units Chinese vermilion similarly dissolved.

(5) Olive oil with 50 per cent. cotton oil yields a tint equal to 5 units Naples yellow, 5 units chrome yellow, and 5 units of vermilion.

(6) Pure cotton oil yields a precipitate having the color of $3\frac{1}{2}$ units chrome yellow, 10 units of vermilion, 1 unit of burnt sienna, and 1 of natural sepia in 100 units of water.

Other seed oils, including sesame, camomile, peanut, and poppy seed, give a precisely similar series of tints in proportion to the degree of their admixture with olive oil, except that the colors are more inclined to the reddish shade which would be produced by covering the corresponding cotton-seed tint with a thin wash of carmine. These gradations of color are most marked when the liquid in the tube is at about the stated temperature, 40° F. As the precipitate is further chilled to the freezing point the colors fade and lose their individuality. Such is the system which is now expected will enable purchasers and consumers of olive oil in this country to detect the adulterations, which have become so general that very few brands or firm names are any longer a guaranty of purity. When it is remembered that more than 2,000,000 gallons of cotton-seed oil are exported from the United States to Marseilles in a single year, and that more than half of this vast quantity is used for adulterating olive oils, a large part of which are re-imported to the United States through a 30 per cent. duty, the importance of some new and better means of controlling the integrity of this trade will be apparent. Some time ago 1,000 tierces of American lard were stopped at the wharf in Marseilles, and the consignees subjected to a costly process, which is not yet terminated, because the lard was found upon analysis by the customs officers to contain 10 per cent. of cottonseed oil. This seizure was based upon the fact that, while lard is entitled to entry duty free, cottonseed oil bears a duty of 6 francs per 100 kilograms, and this adulteration of a free article with a dutiable one is held to be fraudulent. The least that can happen to the shippers in this case will be that they must pay the duty on 100 tierces of cottonseed oil and the expenses of the process, besides the loss which the consignee suffers from the delay. Might not this rigid scrutiny be equally well applied to some of the adulterated and falsified foreign products which are landed at American ports?

It is not within the scope of this report to consider whether either lard or olive oil, when adulterated with cottonseed, is necessarily unwholesome. The vital fact is that in paying from 40 to 50 cents per kilogram and 30 per cent. duty on American cottonseed as olive oil, the people of the United States are submitting to a wholesale fraud, the proportions of which are increasing year by year.

The interest of both the United States and France will be subserved when the reckless tampering with the integrity of commerce is systematically suppressed. As long as our people will accept and pay for adulterated oils they will continue to flood and dominate the market. The remedy must be applied at our ports of entry.

Mr. David Wesson* makes the following comments in regard to Brullé's and other methods of testing for cotton seed oil:

"We have worked some with the chloride of gold test and find it will give a reduction with cottonseed oil, free fatty acids, and old rancid lard. It gives no reduction with pure fresh lard containing less than 1 per cent. of free acid.

"We find the Brullé test is unaffected by free acid or rancidity. We have tried the Bechi test on some highly oxydized cotton oil and find it gives no reduction whatever; while with lard oil made from old lard considerable reduction can be obtained."

COCOA-NUT OIL AS AN ADULTERANT OF LARD.

It is probable that in this country lard is never adulterated with cocoa-nut oil for commercial purposes. Allen† speaks of the use of cocoa-nut oil as an adulterant of lard. In "The Analyst," October, 1888, page 89, he says he is unable to trace the authority on which the statement was made. He has, however, in his own experience found one sample of lard which was adulterated with cocoa-nut oil. This lard gave the following numbers on analysis:

Specific gravity at 99°	.866
Iodine absorption, per cent	37.4
Saponification equivalent	265.2
$\frac{N}{10}$ alkali for the distillate from 2½ grams	3.3cc.

The volatile acids contained a notable proportion of soluble acids of sparing solubility in water, and had the characteristic odor of the distillate from cocoa-nut oil. The sample was certified to contain 33 per cent. of the adulterant. The most accurate determination of the cocoa-nut oil is obtained from the saponification equivalent. Mr. Allen gives

---

*Letter of March 4, 1889.
†Commercial Organic Analysis, Vol. 2, p. 142.

the saponification equivalent of lard at 289 and cocoa-nut oil at 219; hence every .7 fall in the equivalent below 289 indicates the probable presence of 1 per cent. of the adulterant. Pressing inquiries have been sent to Mr. Allen from America as to where cocoa-nut stearine could be obtained, but none was found to be on the market. The comparison of the analyses of pure lard and cocoa-nut oil is given in the following table:

	Lard.	Cocoanut oil.
Original fat:		
Plummet gravity at 99° C	.860 to .861	.868 to .871
Iodine absorption	55 to 61	9
Saponification equivalent	286 to 292	269 to 222
Volume of $\frac{N}{10}$ alkali required by distillate from 5 grams	0.5	7.0
Separated fatty acids:		
Plummet gravity at 99° C	.838 to .840	.844
Iodine absorption	61 to 64	15.61
Mean combining weight	278	200

DETECTION OF COTTONSEED OIL IN LARD.

Mr. A. H. Allen[*] has made a further study of the detection of cottonseed oil in lard. As a result of his analyses he gives the following figures:

	Omentum lard.	Leaf lard.	English lard.	American lard containing cottonseed oil.	Mixture of unknown nature.	Suspected sample.
Original fat:						
Melting point, C°	39.0	40.0	39.0	37.5	40.0	....
Solidifying point, C°	36.5	32.0	27.0	27.5	30.5	....
Plummet gravity at 99 C°	.8602	.8620	.8203	.8618	.8637	.8637
Iodine absorption, per cent	55.4	60.5	64.0	82.5	68.8	62.8
Fatty acids:						
Melting point, C°	39.0	39.5	....	....	39.5	....
Solidifying point, C°	28.7	38.5	....	....	37.5	....
Plummet gravity at 99 C°	.8372	.8385	....	....	.8450	.8385
Mean combining weight	274.5	....	....	....	276.8	....
Iodine absorption, per cent	58.3	65.3	....	....	70.4	64.8
Oleic acid, etc., per cent	58.4	....	....	....	57.8	....
Oleic acid, iodine absorption	87.4	....	....	....	(94.6)	....
Millian's nitrate of silver test	White.	White.	Gray.	Marked blackening.	Marked blackening.	Sensible darkening.

[1] Rising to 27.5   [2] Rising to 39.0   [3] Rising to 38.8   [4] Rising to 38.5

*The Analyst, September, 1888.

He also gives the result of a comparison of tallow, lard, and cotton oil:

	Tallow.	Lard.	Cotton oil.
Original fat:			
Melting point, C°		38 to 45	Fluid
Solidifying point, C°	33 to 48		
Specific gravity at 99 C°	.862	.860 to .861	.872
Iodine absorption, per cent	40	59 to 62	105 to 110
Fatty acids:			
Melting point, C°	45	38	35 to 36
Solidifying point, C°	43	38	30
Specific gravity at 99 C°			.8467
Iodine absorption, per cent	41.3	61.2	115.7

Analyses were also made of cotton oil and cotton-oil acids, as indicated in the following table:

	A.—Cotton oil stearine	B.—Cotton oil.	C.—Cotton-oil acids from B.
Pinnacot gravity at 99 C°	.8684	.725	.8476
Melting point C°	40		35
Solidifying point C°	*31		32
Iodine absorption	59.8	108 to 110	115.8
Saponification equivalent		285 to 294	289
Acidity (=oleic acid)	.94	Trace.	97.6

* Rising to 32.5°.

It is found that there is a marked difference in the specific gravity of lard and cotton oil, and also in the iodine absorption of the two. Lard and beef fat have substantially the same specific gravity. The difference is important, since it would enable one to distinguish a mixture of beef stearine and cotton oil, having an iodine absorption of about 60 from genuine lard. Thus, with a proportion of the adulterant in a mixed composition of lard, the cotton oil only can be ascertained with considerable accuracy by determining the iodine absorption; the estimation will be below the truth if beef stearine be present. On the other hand, the presence of beef stearine does not interfere with the deduction to be drawn from the increased specific gravity of the melted sample.

Mr. Allen finds Milliau's nitrate of silver test to be valuable, and prefers it to the original one proposed by Bechi. In his opinion the indications obtained from the melting-point or solidifying point of the glycerides of the fatty acids are of no value. Samples of lard oil were found to have an iodine absorption of 73 and 74, while one several years old gave only 41. It is recommended that the iodine absorption be determined on the fatty acids instead of the original glycerides, thus avoiding the use of chloroform, which has a marked disturbing influence on the strength of the iodine solution employed.

Hehner[*] states that in Bechi's test, without impairment of the delicacy, the re-agent may be made up without the amyl alcohol or rape-seed oil. He makes the solution of nitrate of silver in alcohol and ether very slightly acidified, and adds to the oil to be examined about one-half of the bulk of the silver solution, and then heats on the water-bath for one-quarter of an hour longer. Pure lard always remains perfectly unchanged by this treatment, while cotton-oil mixtures blacken more or less quickly. It is quite possible to arrive at approximate quantitative results by comparing the oil mixtures of known composition. Mr. Hehner does not see any advantage in Millian's modification. The rise of temperature when mixed with sulphuric acid is to be preferred as a method of estimating the quantity of cotton oil in lard. The sample, of course, must be free from water. When 50 grams of pure lard, according to Hehner, are mixed with 10cc of strong sulphuric acid the rise of temperature varies from 24° to 27.5°, while cotton oil in the same conditions shows an increase of 70°. In every case lard which reduces silver shows an increase of temperature of more than 27.5°.

[NOTE.—Compare these temperatures with those obtained in our experiments. The mean rise of temperature for pure lard was 41.5°, and the mean increase for cotton oil 85.4°.]

Roland Williams[†] has also contributed a study to the adulteration of lard with cotton oil. He regards the saponification equivalent as quite useless as far as the detection of cotton oil in lard is concerned, as both the lard and cotton oil require practically the same amount of alkali for saponification. In case of the use of cocoa-nut oil, however, the determination of the saponification equivalent is of the highest importance.

The melting-point also is regarded as of no value in respect of the detection of adulteration, since it depends largely on the parts of the animal from which the fat has been obtained. The specific gravity of pure lard at the boiling-point of water is about .861, and of cotton-oil at the same temperature .872. It may be possible, therefore, to derive some valuable information in regard to the constitution of lard or mixed lards from a careful determination of the specific gravity. Mr. Williams failed to obtain valuable results with Maumené's test. This failure was doubtless due to some imperfection in the method of manipulation.

In the absence of interfering bodies Mr. Williams relies chiefly upon the percentage of iodine absorbed in estimating approximately the amount of cotton oil present as an adulterant. The addition of stearine to lard interferes seriously with the determination of the percentage of added cotton oil by the iodine method. He has found pure lards to absorb from 60 to 62 per cent. of iodine. One sample of lard, said to be leaf lard, absorbed only 51 per cent. Some leaf lard rendered by Mr. Allen himself absorbed 51.8 per cent.

---

[*] The Analyst, September 1888, p. 165, *et seq*.
[†] The Analyst, September, 1888, pp. 168, 169.

Milliau's modification is recommended, but it is advised that a blank experiment be made with the re-agent, since sometimes alcohol contains impurities which reduce silver nitrate. Experiments in the use of the silver nitrate test for quantitative purposes did not give satisfactory results.

Jones* says that he was the first public analyst of England to certify a case of lard adulterated with cotton oil under the sale of food and drugs act. He first applied a qualitative test with chloride of sulphur essentially the process described by Warren. He used 5 grams of the fluid lard in a porcelain dish, to which he adds 2cc of equal volumes of chloride of sulphur and bisulphide of carbon. The mixture is well stirred at first and occasionally for fifteen or twenty minutes. No heat is applied. By this treatment genuine lard only thickens or becomes rather stiff in three hours. If it contain cotton oil it becomes quite hard and solid in one-half hour. This test is very simple, but with practice one can with certainty pick out all lards containing cotton oil. He estimates the extent of the adulteration by the percentage of iodine absorbed. He finds that pure lard never takes sensibly more than 60 per cent. of iodine, while cotton oil takes 105 to 110 per cent. He adopts the formula—

$$100 \left(\frac{\text{I. absorbed} - 60}{45} \right) = \text{per cent. cotton oil.}$$

The accuracy of the work is checked by the specific gravity taken at 100° F. At this temperature the specific gravity of pure lard is taken at .9060, and of cotton oil at .9135.

The formula for calculating the percentage of adulteration by the specific gravity is as follows:

$$100 \left(\frac{\text{Sp. gr. found} - 906}{7.5} \right) = \text{per cent. cotton oil.}$$

The radical error in the method of Mr. Jones is, that he takes no account whatever of the admixture of stearine with adulterated lard, which may be done so skillfully as to wholly vitiate the method employed for determining the amount of adulteration.

Stock† describes a modification of Milliau's method for the detection of cotton oil. His method is as follows:

Fifteen grams of the sample are saponified in a 7 inch porcelain basin with a mixture of 15cc of 30 per cent. NaHO and 15cc of 92 per cent. alcohol. To commence, the fat is heated to 110° C. The alkaline alcohol must be added in quantities not exceeding 1cc at a time, the

---

*The Analyst, September 1888, p. 170.
†The Analyst, September 1888, p. 172.

temperature not being allowed to fall below 95° C. to 100° C., constant stirring at this part of the operation being most important. If the saponification has been successful, the resultant soap is a smooth, thick paste. Boiling distilled water is now added drop by drop, a thin, flexible spatula being used to break down the paste. When this has the appearance of smooth starch, water may be run in till a volume of 500cc is reached. Complete solution should follow. Forty cubic centimeters of diluted sulphuric acid (1—10) are now added to the contents of the basin, the liquid is stirred gently and brought to boil for seven to twelve minutes, then kept just below boiling, until the separated fatty acids fuse to a clear oily layer. The greater bulk of the acid watery liquid is siphoned off, the remainder with the fatty acids being poured into a clean, warm flask with a somewhat long and narrow neck. The fatty acids are freed as nearly as possible by siphonage from the watery under layer, and the flask is filled up with boiling water so as to bring the fatty acids into the neck, by which operation a partial washing is given. Five cubic centimeters of the fused fatty acids are now transferred by means of a dry, warm, fast-running pipette, into a clean, dry, wide test tube. Twenty cubic centimeters of absolute alcohol are added, care being taken to wash the pipette by running the alcohol through it. The contents of the test tube are heated to incipient ebullition in a vessel of boiling water. Two cubic centimeters of a 50 per cent. solution of silver nitrate are now rapidly poured into the tube, when, if even 2 per cent. of cotton oil be present in the sample, the characteristic cedar-brown color is at once developed. Pure lard gives absolutely no color.

To quantify this reaction, known mixtures of pure lard and refined cotton oil are treated exactly as above, and the colors in the different tubes compared by reflected light against a white background. This must be done simultaneously, for in about seven minutes the coloring matter begins to fall out, and correct comparison is then impossible. In careful hands excellent results are obtainable.

Prof. J. Campbell Brown[*] calls attention to errors analysts are liable to make:

1. They are liable to underestimate the proportion of cotton oil when relying upon the iodine test alone. The reason of this is found in the admixture of stearine in adulterated lards which has a low iodine number.

2. They are liable to condemn genuine lard which is more oily than pork fat or lard rendered in England. According to Hehner American lard contains more olein than English. I do not think the assumption of Hehner a just one since the iodine number of pure lards in this country is found to be about the same as in England.

Mr. Watson Grey[†] gives the results of some determinations of the

---

[*] Op. cit., p. 103. [†] Op. cit.

absorption of iodine by lard showing a very low absorbtive power. His results are given in the following table:

Kind of lard.	Iodine absorption.
	Per cent.
From omentum of hog	49.5
Market lard (bought in Liverpool)	49
From omentum of sow	55.5
From back of pig	65

Mr. Grey will fix the average for English lards at 57 per cent. instead of 62 as taken by Mr. Allen. Mr. Fox stated that he had recently found 50 per cent. of pea-nut oil in lard oil, determining it by the altered specific gravity and the presence of arachidic acid.

Mr. M. F. Horn* gives a method for the quantitative estimate of paraffine, cerosin and mineral oils, in fats and wax. Inasmuch as these adulterations are not likely to occur in lard I will cite only the original paper.

Roland Williams† gives a table showing the iodine numbers and melting-points of certain fatty acids. The melting-points were determined by the ordinary capillary-tube method. Following are his results:

Name of fatty acid.	Iodine absorption.	Melting-point.
	Per cent.	°F.
Tallow	41.3	119
Lard	61.2	100
Cotton oil	115.7	96
Olive oil	90.2	81
Linseed oil	178.5	75
Rape oil	105.6	71
Castor oil	83.9	
Cocoa-nut oil	9.5	75
Palm oil	55.4	111
Sperm oil	88.1	56

The low melting-point in the case of lards is explained by Mr. Williams on account of the fat having been taken from the entire animal. As might be expected the fatty acids absorb a slightly greater percentage of iodine than the glycerides from which they were made.

Prof. Stephen P. Sharpless‡ relies upon the usual tests for the detection of the adulteration of lard with cotton oil. Bechi's test, he says, gives good results. Nitric acid of 1.35 specific gravity gives only a faint color with pure lard, while with lard adulterated with cotton oil it gives a color more or less intense. For the detection of added stearine made from tallow Dr. Belfield's microscopic test is employed. The suspected

* The Analyst, Oct. 1888, p. 184. Zeitschr. f. Angew. Chemie, No. 16, 1888, p. 459.
† The Analyst, May, 1888, p. 88.
‡ The Analyst, April, 1888, p. 69.

lard is dissolved in ether in a test tube, which should be about two-thirds filled. The solution should be nearly saturated. The tube is loosely stopped with cotton wool, and placed in a quiet room, at a temperature of about 60° F. When the first crystals are formed they are removed by means of a pipette, placed on the slide of the microscope, and examined in the usual way. The forms of the crystals produced have already been described.

David Wesson* says of Belfield's microscopic test, that while at times it gives very characteristic crystals, at other times their forms are not sufficiently definite to be relied upon. The nitric and sulphuric acid tests are sometimes unreliable, especially with old samples. Bechi's test is also sometimes uncertain. On old samples of cotton oil it sometimes gives negative results, while with old samples of lard oil it will give a slight reduction.

Michael Conroy† uses the following tests for the determination of the purity of a sample of lard.

(1) Heat and stir about one-half ounce of lard with one-tenth its weight of strong nitric acid, specific gravity 1.42, in a porcelain dish of about 8 ounces capacity, until a brisk action commences, when the source of heat should be removed. Pure lard sets in about one hour to a pale orange-colored solid, but if it contain cotton oil it takes a more or less deep orange-brown tint.

(2) The test of Labiche was also tried, as follows: Equal parts of the fat and neutral acetate of lead and ammonia added, stirring briskly. The acetate of lead decomposes and the nascent oxide reacts upon the oil, causing it to turn red. This reaction proved a failure.

(3) The proceeding of Ernest Milliau By this test it is claimed 1 per cent. of cotton oil can be detected.

(4) Bechi's test: When sodium carbonate has been used to correct the acidity of lard this test is not applicable, unless the re agent be acidified with nitric acid. The following modification of Bechi's test was employed: A solution of five parts of silver nitrate and one part of nitric acid, specific gravity 1.42, in one hundred parts of alcohol. Put 6 grams of lard in a dry test tube and add one-fourth gram of the solution above described, and hold the tube in boiling water for five minutes. Pure lard remains perfectly white, but if adulterated with cotton oil it assumes a more or less olive-brown color. This color is best observed when the lard sets. One per cent. of cotton oil in a lard gave a color quite distinct from the genuine article.

Cotton oil has also been used for the adulteration of tallow.‡

The melting-point of the genuine tallow, according to Williams, varies considerably in different samples, ranging from 100° to 120° F.

---

* The Analyst, July, 1888, p. 110.

† The Analyst, Vol. 13, No. 151, p. 203. The Pharmaceutical Journal and Transactions, September 22, 1888, p. 237.

‡ Roland Williams, Journal of Society of Chemical Industry, March, 1888, p. 186.

The best class of tallow has a melting-point of about 110° F. Pure tallow requires from 19.3 per cent. to 19.8 per cent. of caustic potash for saponification, and cotton oil 19.1 to 19.6. A series of mixtures of tallow and cotton oil was prepared containing 5, 10, 15, 20, 25, 30, and 40 per cent. of the oil. The addition of the cotton oil did not have the effect upon the melting-point which might be expected. The pure samples melted at 110° F. and the one with 40 per cent. oil at 102° F. The quantity of iodine absorbed was by the pure tallow 40.8 per cent., and by the mixture containing 40 per cent. oil 66.2 per cent. The percentages for the several samples were as follows: 44, 47.1, 49.7, 52.9, 56.1, 59.2, 66.2. The percentage of iodine absorbed by the original cotton oil was 109.1 per cent. The percentages of iodine absorption have a remarkably close connection with the percentage of cotton oil present in the various mixtures.

REACTION OF ANIMAL AND VEGETABLE GLYCERIDES FOR CHOLESTERIN AND PHYTOSTERIN.

The presence of cholesterin in animal glycerides, especially liver fat, has long been known.

A substance homologous with cholesterin was detected in the oil of Calabar beans in 1878 by Hesse, to which he gave the name of phytosterin.[*]

Salkowski proposes to distinguish animal and vegetable fats from each other by testing them for cholesterin and phytosterin respectively[†]. To obtain the cholesterin (phytosterin) 50 grams of the glycerides, animal or vegetable are saponified with alcoholic potash. The alcohol is evaporated, and the soap diluted with water to about 2 liters. This is shaken in a separating globe with ether, and the ether solution drawn off and evaporated to small bulk. The residue, which may contain a small quantity of unsaponified fat, is again treated with potash, and the separation effected by ether, as above, only a little water being added. If the ether solutions separate slowly, a few drops of alcohol may be added.

The ethereal extract is evaporated and the cholesterin separated in crystals. Animal cholesterin has a melting-point of 146°; vegetable (phytosterin) 132°. The two also show distinctly crystalline forms which are easilily distinguished under the miscroscope. Vegetable cholesterin shows star-shaped crystals or bundles of long, quite solid, needles, while the animal product gives thin rhombic tables.

Dissolved in chloroform, the two products show different color reactions with strong sulphuric acid. Cholesterin shows a cherry-red and phytosterin a blue-red color. In mixtures of animal and vegetable glycerides the melting point of the cholesterin obtained may become a fair index of the proportion of the two present. Thus, a melting-point of 139°

---

[*] An. d. Chem. u. Pharm., Vol. 192, p. 178.
[†] Zeitschrift für Analytische Chemie, Vol. 26, p. 572.

would indicate that the fat from which the cholesterin was obtained was made up of equal proportions of animal and vegetable glycerides.

SEPARATION OF STEARINE AND PALMITINE IN LARD.

Isbert and Venator* have separated stearine and palmitine from lard in the following manner:

The sample is dissolved in cold ether in a test tube, and the closed tube allowed to stand for some time. At the end of about two hours the stearine begins to separate and is collected at the bottom of the tube. The identity of the stearine was shown by its melting-point, viz, 60°. The palmitin separates later.

The separation can also be effected by solution in boiling alcohol. The separated glycerides are separated from olein by pressing between blotting paper.

ABSORPTION OF OXYGEN.

Cotton oil absorbs a notable proportion of oxygen when subjected to the Livache process.†

Finely-divided lead is obtained by precipitating with zinc. About 1 gram of the lead powder is placed on a watch glass and mixed with nearly 5 grams of oil. The disk is placed in a well-lighted room of medium temperature.

Cotton oil gains about 6 per cent. in weight in forty-eight hours. The equivalent of oxygen absorption may also be approximately calculated for cotton oil from its iodine number by multiplying this by .063 ($\frac{1\cdot 1}{2\cdot 1} = .063$).

For cotton oil the number thus obtained is 6.7 per cent.

ELAIDINE REACTION.

Oleic acid under the influence of nitrous acid is converted into an isomeric elaidic acid.

In like manner triolein $C_3H_5(OC_{18}H_{33}O)_3$ is converted into elaidine. This substance is formed in crystalline masses, and its melting-point is variously given at 32° to 38°. Following is the method of applying the elaidine test known as Pontet's process in the municipal laboratory of Paris:

	Grams.
Of the oil to be tested	10
Nitric acid	5
Mercury	1

Place in a test tube and shake vigorously for three minutes until the mercury is dissolved; allow to stand for twenty minutes, and shake again for one minute.

In from one to three hours the sample becomes hard. Olive, pea-nut, and lard oils give the hardest elaidines. Copper may be used instead

---

* Zeit. f. Angew. Chemie, June 1, 1888, p. 316.
† Moniteur Scientifique, Vol. 12, p. 253 et seq.

of mercury, in which case the nitric acid should be somewhat diluted. The red vapors produced by the action of iron on nitric acid may also be conducted directly into the oil.

One part of the strong nitric acid may also be shaken with three to five parts of the oil and a solution of nitrite of potash added drop by drop with constant shaking.

Attempts have been made to measure the relative hardness of the elaidine produced by the distance which a plunger carrying a known weight would sink into it, and the data thus obtained have been used for quantitative calculations.

SPECTROSCOPIC EXAMINATION.

The absorption spectrum of an oil depends upon the character of the coloring matter contained therein. Many vegetable oils give a spectrum characteristic of chlorophyll.

Cotton oil gives a banded absorption spectrum.

The use of the spectroscope in examinations for lard adulteration is probably not as extensive and general as the merits of the process would warrant.

FURTHER QUALITATIVE REACTIONS.

There are other qualitative reactions which might sometimes prove of value in the examination of lard and its adulterations.

These are the methods of Chateau, Fauré, Heydenreich, Penot, Crace Calvert, Flückiger, and Glaessner.

A full description of these methods is given by Benedikt.*

ABSTRACTS OF METHODS OF LARD ANALYSES, WITH RESULTS THEREOF.

[Employed in the case of McGeoch, Everingham & Co. vs. Fowler Bros., before Chicago Board of Trade.]

Much progress has been made in the science of lard analysis since the famous case of McGeoch, Everingham & Co. vs. Fowler Bros., the notes of which have been published in pamphlet form by Knight & Leonard, Chicago, 1883.

The complaint against the Fowler Bros. rested on the charge that they had sold prime steam lard which contained other than hog fat. The complaint was brought before the Chicago Board of Trade, and was heard by the board of directors thereof. Samples of the suspected lard were submitted to a large number of chemists, and an abstract of their methods of analyses and the results obtained follows:

TESTIMONY OF DR. P. B. ROSE.†

He can generally tell, when a sample of prime steam lard is sent to him, if there have been any impurities put into it, by examining its color and quality; the samples are sent to him for the purpose of seeing whether the lard is up to the proper

---

* Analyse der Fette und Wachsarten, p. 198, et seq.
† Pamphlet mentioned, p. 116.

standard, whether it is off-color, or anything of that sort; sometimes lard is of too dark a color; a small quantity of tallow in lard could not be detected by its appearance to the naked eye; a thousand or twelve hundred pounds of tallow put into one or two tanks could not be detected by the eye; he thinks during last November only a thousand or twelve hundred pounds of tallow was received into the house from all sources. Tallow fat is worth 7½ to 8 cents per pound; he has never tried it, and does not know how much tallow could be put in a tank of lard without it being detected. 25 or 20 per cent. could be detected, and he thinks 15 per cent. could be readily detected by the naked eye and by the taste; he has never tried 10 per cent; he thinks an inspector would readily detect 15 per cent., and with 10 per cent. of tallow he thinks an inspector would discover there was something wrong.

TESTIMONY OF PROF. M. DELAFONTAINE.*

CHICAGO, *June* 6, 1883.

To whom it may concern:

This is to certify that on or about the 22d of May and the 2d of June, 1883, I received from Mr. Mixer, provision inspector of the Board of Trade, three samples of lard, respectively marked 1, 2, and 3. Mr. P. McGeoch requested me to analyze them, and I find that neither of them is pure hog fat. Samples Nos. 1 and 2 gave indications of cotton-seed oil, and both contain a percentage of beef stearine (or a corresponding quantity of beef tallow) exceeding 10 per cent. Owing to the smallness of the quantity of No. 3 at my disposal, I can not testify positively and beyond reasonable doubt about the presence or absence of cotton-seed oil, but the proportion of beef stearine is at least equal to that found in the other samples.

M. DELAFONTAINE.

The experiments were all comparative; the same weight of each substance and the same bulk of solvents used, drawn from the same supply; the vessels were of the same kind and capacity; the experiments were conducted on the same table, at the same window, etc.; nothing was different but the final results for different samples; temperature between 12 and 15 degrees centigrade. For the detection of cotton-seed oil olein was extracted, as usual, and tested by the elaidine test (the taste and smell were noted too). For the extra stearine the lard was treated with eight or nine times its weight of pure alcohol and ether, half and half, allowed to stand twenty-four hours, liquid then poured out and replaced by a little over half as much again of the solvent, shaken often, filtered after eight or ten hours, dried, weighed.

First. What quantity of lard did you operate upon?
Answer. For some experiments on about $4\frac{7}{10}$ grams; for others on twice that quantity; for others on 20 grams.
Second. What was the liquid you used to dissolve the lard in? if a mixture, state what proportions of each liquid.
Answer. Half Squibb's ether, and half Squibb's absolute alcohol.

* * * * * * *

Fifth. Did you heat the lard and add to it the mixture, or did you simply add the lard and then apply the heat?
Answer. The lard was heated to about 70 degrees C.

* * * * * * *

Seventh. What was the shape and size of the vessel in which you thus treated it?
Answer. Cylindrical glass jars, glass-stoppered, graduated, holding 50, 100, and 200cc.
Eighth. Did you decant the liquid off?
Answer. I did.

* * * * * * *

---

*Op. cit., pp. 139, 140, 141, 142, 143.

Tenth. After filtering or decanting as above, how did you treat it, or where did you keep it before weighing?

Answer. Dried it in an air-bath.

Eleventh. Did you weigh the residue while on a filter or in a beaker, or evaporating dish, or how?

Answer. On the filter.

Twelfth. What was the exact weight of the residue found?

Answer. I have kept a record only of the results; the only figures that I can find just now of the actual weight of the residue are the following:

    4.25 grams of lard, No. 3, gave 275 milligrams.
    4.7 grams of pure lard gave 350 milligrams.
    19.74 grams of pure lard gave 200 milligrams.
    9.4 grams of lard, No. 3, gave 600 milligrams.

In making his sample test of pure lard he took his material chiefly from the leaf lard and from the sides of the hog; some of it was salted and some was fresh; before rendering it he cut the material into very small pieces, and allowed it to stand in a large volume of cold water for some time to take the salt out; it was then filtered out in pans and rendered on a sand-bath, that is, pans full of sand and heated from below, so as to get an even temperature and not burn the lard; after rendering, the lard was filtered, in order to remove any tissue or foreign matter from it; he is quite sure he got, into 1 per cent., all the lard there was in the material; the temperature at which the lard was rendered was 175 to 200 degrees centigrade,* which is much higher than is necessary to break up the cells and melt all the stearine there may be in the lard. In getting the samples of lard from a packing-house he asked for pure prime steam lard; in testing that sample he found it to run a little higher in stearine than the lard he rendered himself; he can not, of course, say that the sample procured from the packing-house was perfectly pure, because he did not himself see it rendered. The solvent he used was absolute alcohol and the strongest of Squibb's ether; he always measured the solvent; the melted lard was at about 70 or 75 degrees centigrade, when the solvent was applied, so as to be sure the palmitine would remain in solution. After the solvent was mixed with the lard he did not ascertain its temperature; he had no use for that; the lard was allowed to stand in the solution before decanting about fifteen to eighteen and sometimes twenty-four hours, during which time it was kept in cold water, at a temperature 12 to 15 degrees centigrade, always below 15 and sometimes a little lower than 12; he does not know what the temperature of the room was in which the mixture was allowed to remain. The mixture of alcohol and ether, after being added to the lard, was well shaken; after the first solvent had been decanted, he replaced it with about half as much of fresh solvent as had been first used, shook it well and often for two or three hours, and then allowed it to stand ten or twelve hours or so, sometimes over night, again shook it several times, then filtered, and in order to avoid an error that might arise from the liquid evaporating and leaving a part of the residue too hard, he pressed it between blotting papers so as to absorb all that was not properly residue, then dried in an oven and weighed it; sometimes, while it was on the filter, he poured more of the solvent on it and again filtered so as to dispose of all that was soluble. He can not see what the residue could contain except stearine, unless it might be a small quantity of palmitine; he tested the residue by determining its melting point and its solubility, and that showed it to be stearine; there is no difference in the chemical characteristics of the pure stearine procured from the fat of beef, mutton, or pork; they are the same thing as far as he knows, and he does not know of any difference in the chemical reaction of these different kinds of stearine; he does not attempt to distinguish between them; he does not know certainly how much pure stearine lard actually contains; it varies; he has never found it to exceed 2 per cent. in pure lard, and sometimes it runs as low as three-quarters of 1 per cent. when subjected to the

---
*Probably Fahrenheit is meant.

process for extracting it he has described; very likely it would vary that much in lard made from different parts of the same hog; he speaks on these points from his own experiments; he has not looked for authorities on this subject; he is now performing some experiments which he hopes will throw some light on this branch of the subject.

* * * * * * *

In testing for cotton-seed oil he extracted the olein by means of absolute alcohol, heated, allowing the liquid to cool and then filtering and drying off the alcohol; he takes a glass flask or anything capable of holding the lard and pours over it some absolute alcohol, and boils the two together for a few minutes, then allows the mixture to cool; this produces a crystallization; and then having kept it cold for a number of hours he filters it, and the liquid is for all practical purposes a solution of olein and alcohol; the alcohol is then driven off from it and what is left is olein. In the case of these lard samples he treated the olein by the elaidine test, using as a liquid sulphuric acid saturated with the red fumes of hyponitric acid; by this treatment the olein of oils is turned into a hard solid mass; olein is naturally a liquid, but when the test is applied to cotton-seed oil the oil remains floating. The same test applied to pure lard oil or pure olive oil soon turns the oil hard. If cotton-seed oil and lard oil are mixed with this liquid the mixture will solidify only after a much longer time than would be required to solidify pure lard oil, or often it will not solidify at all, depending upon the proportion of the cotton-seed oil; he took a glass test-tube and put into it a certain quantity of the olein to be tested, and the acid to about half the bulk of the olein, shook it well, and kept the tube at a temperature of about 10 degrees centigrade; he observed the time it took for the liquor to solidify. Nitric acid for use in the elaidine test is not reliable and he did not use it; he depends upon the absence of solidification after half an hour to determine the basis of reaction. Lard oil solidifies pretty quickly when treated by the acid he employed; cotton-seed oil does not for several hours.

He does not know of any writer who has stated that cotton-seed oil can not be detected when it is present in less proportion than 5 to 10 per cent.; he has that information from personal conversation with others. His method of analyzing lard is not one published in the books, as far as he knows; he adopts it mainly as the result of putting this and that together. He is not willing to take ten samples of lard prepared by a competent and reliable expert, whose certificate as to what they contain shall be placed in the hands of the president of the Board of Trade and stake his reputation on being able to tell which are adulterated and which pure, using in the analysis the methods he has employed in testing the samples in respect to which he has been testifying, because a mixture can be made with fats or some foreign oil which he has not sufficiently studied to be able to certainly detect such substances; his examinations have been with reference to detecting substances which are most likely to be used for the purpose of adulterations, such as tallow and some other substances. In the case of a mixture of equal proportions of pure lard with a lard from which, say, half of the lard oil has been expressed, leaving the mixture deficient in lard oil to the extent of 25 per cent., that mixture would be found to contain more stearine than pure lard.

EVIDENCE OF WILLIAM HOSKINS.*

CHICAGO, *June* 5, 1883.

This certifies that I have analyzed a sample of lard received from Mr. C. H. S. Mixer, marked No. 3, and find that it is adulterated with at least 20 per cent. of beef stearine or its equivalent of tallow; and further, I find evidences of the presence of cottonseed oil, or one of its derivatives.

G. A. MARINER.
Per HOSKINS.

---

* Op. cit., pp. 147, 148.

CHICAGO, June 1, 1883.

This certifies that I have analyzed two samples of lard, marked respectively No. 1 and No. 2, received from Mr. Mixer on May 30, 1883, and find that both are adulterated with beef stearine or its equivalent amount of tallow to the extent of at least 20 per cent.

G. A. MARINER.
Per HOSKINS.

One of his methods was by taking equal proportions of alcohol and ether, and into that mixture putting a certain amount of pure lard, and also an equal weight of the samples to be tested. The lards were warmed, and were then poured into the vessels containing the mixture under exactly the same conditions. After some time more or less of the stearine separated. In pure lard, treated in the way he has described, the separated substance, which is chiefly stearine, rarely exceeds 1 per cent. One of the samples given him by Mr. Mixer gave less than 4 per cent.; one gave over 5 per cent. Another process, known as Blythe's pattern process, is to take a piece of glass, chemically cleaned, and having a thin film of water on it. On this is dropped a drop of the melted substance. In the case of lard one pattern is produced, in the case of tallow a different pattern, and in the case of the mixture of the two a still different and intermediate pattern is produced. He regards this as an absolute test and one easily applied. Another test used is to ascertain the difference in time taken in saponifying samples. Tallow takes much less time to saponify than lard does, under proper and the same conditions. This process gives quite accurate results. These are the chief tests he depended upon in his chemical examinations of the samples now in question. The processes he has described are recognized by authorities, and have all been published as authority. He has during the past two winters had considerable experience in examinations as to the adulteration of butter, and has studied the subject of fats to a considerable extent. He considers the results of his examination of the samples given him by Mr. Mixer as conclusive in respect to their quality; there is no possibility of a doubt as to the correctness of his conclusions in respect to them. In regard to the presence of foreign oil in lard there is an absolute test, known as the elaidine test. It especially applies to drying oils. Nitrous oxide is made by heating mercury with nitric acid. In treating non-drying oils with it the point noted is the fat which is solid. It makes no such combination with drying oils, and when they are in large quantity they separate and come to the top and can be seen as liquid. In other cases they form more or less of a semi-solid; he thinks almost anybody could see the difference when pointed out, although it requires considerable experience in the use of the microscope to be able to detect these differences unaided. All his examinations were carried on parallel with examinations of samples of pure lard rendered by himself, at least a part of which was leaf lard. He thinks there may be a little difference between the proportions of stearine in leaf lard and fat taken from the sides of the hog. It is, however, but slight. After all his examinations of the samples in question in this case, it is his absolute conclusion that the lard is adulterated to the extent of 20 per cent. of foreign material. He calculates the percentage of adulteration from the basis that, as pure lard never contains more than 1 per cent. of pure stearine, and that tallow contains about 9 per cent., therefore, as these samples contained none less than 4 per cent., there must have been added to the lard beef tallow, or some of its derivatives, sufficient to account for the 3 per cent. of excess, and as three nines are twenty-seven, he concludes there was over 20 per cent. of adulteration in the samples. The test of adulteration is by the amount of pure stearine found in the sample.

EXACT METHOD OF ANALYSIS.*

A portion of the fat was warmed and mixed with about ten times its weight of a mixture of equal parts of absolute alcohol and ether. After allowing it to stand about twenty-four hours the residue was filtered and weighed.

*Op. cit., p. 119.

A plate of chemically clean glass was covered with a film of water, a drop of the melted fat was dropped upon the plate, and the patterns noted.

A portion of the fat was saponified with an equal weight of sodic hydrate dissolved in water, and the time occupied in saponifying noted.

A portion of the fat was treated with a solution of hyponitric acid and sulphuric acid, and the time necessary for the solidification of the elaidine noted.

In all the above parallel experiments were made under exactly the same circumstances in every respect.

W. HOSKINS.

The above methods rests chiefly on the percentage of insoluble residue after treating the fat with a mixture of ether and alcohol as described.

METHOD OF J. M. HIRSH.*

I melt the sample and dissolve it in purified naphtha, leaving it there at rest at a temperature of about 70° F. for twelve hours, when added stearine or tallow will deposit, while pure lard will show no deposit, or barely a trace. The amount of the deposit increases considerably in the next twelve hours in a mixture with stearine, but little in pure lard. The test being made in a graduated tube, the proportions can be read off without possible error by washing, weighing, or measuring. After the time mentioned the solids of the lard deposit. The remaining solution I treat with nitric acid, which renders crystalline the animal oil (producing elaidine), but leaves the cotton-seed oil a colored liquid.

As a rule the melted sample has minute fibers of cotton floating when it is contaminated with cotton-seed oil; this test is simple and infallible; for this reason I omit to mention other corroborative tests.

J. M. HIRSH.

FURTHER REMARKS BY J. M. HIRSH.†

Lard will give a reaction of elaidine as well as cotton-seed oil, but a time amply sufficient to make elaidine from lard or any animal oil must be greatly exceeded to get the same result from cotton-seed oil. If linseed oil or cotton-seed oil, or the two mixed, are boiled for five minutes with a fume of nitric acid, there will be no apparent change except that they become colorless; they will have to boil an hour or two before separation takes place; in half an hour or so they would become a solid, like stearine. Animal oils treated in the same manner will be solidified and converted into elaidine in five minutes. In applying the elaidine test he first took all the crystals out of the solution, then drew off from the tube all the olein and the benzine, put the nitric acid into that liquid, and heated it; the benzine evaporated quickly; the heating was continued for a few minutes longer, and then it was allowed to cool; the crystallized deposit of elaidine he considered as from animal oils, and what was left, after withdrawing the liquid, he considered was from an admixture of some other oil. In a lard rendered at a high pressure of steam there would be a greater amount of stearine than in one rendered at a low pressure, the original material being the same. If lard is rendered and run into a large reservoir holding, say, 250 tierces, and there allowed to stand for a time, the lard from the bottom of that reservoir would contain a greater amount of stearine than would that drawn from the top. There would be a great deal of difference. The heaviest lard would settle in the reservoir. The difference in the stearine would not be more than, if so much as, 2 to 4 per cent. of chemically pure stearine. A good deal would depend on the depth of the tank or reservoir and on the temperature maintained in the lard. H₃ thinks in lard drawn from the top or bottom of such a reservoir there would not be as much difference in the stearine as he has found in the lard delivered him by

---

* Op. cit., p. 154. † Op. cit., p. 155.

Professor Delafontaine over what pure lard should contain. There is some difference in lard on account of the season at which the hogs are killed, and on account of the age, feed, and other conditions of the animals from which the lard is made.

METHOD OF PROF. C. B. GIBSON.*

Five cubic centimeters of the molten sample, at a temperature of 90° to 100° C., were dissolved into 45cc of half solution of absolute alcohol and ether; this mixture was allowed to stand eight hours, at a temperature of 5° to 10° C., and the stearine allowed to crystallize out; the supernatant liquid was then poured off and 25cc of the fresh solution added to dissolve any remaining olein and palmitin, and allowed to stand twelve hours at 5° to 10° C. The liquid was then filtered off and the residue collected on a tared filter; this was washed until no more fat globules were deposited on evaporating a drop of the washings (alcohol and ether solution); the filter and contents were then dried at a temperature of 30° to 40° C., and by means of desiccator, weighed and the result calculated. All samples were treated at the same time and under the same conditions. In the test for cottonseed oil, I submitted 9cc of the molten lard to the action of sulphuric acid, saturated with nitrous and nitric anhydride. 7cc; the test kept at a temperature of 5° C., until solidification took place, which, in the case of the presence of cotton-seed oil, is produced only after long time; from these results I drew my conclusions. All samples were tested at same time and under same conditions.

C. B. GIBSON.

REMARKS BY PROF. C. B. GIBSON.†

He took the two samples that he received from Professor Delafontaine, a sample he prepared himself, and a sample he procured in the market, said to be absolutely pure lard, and treated them all by the methods he has described in his written statement; all the samples were treated in corked tubes; the samples all produced different results, some considerably different, and in the two samples he rendered himself there was a slight variation; he tested the process by comparison with the pure lard he had rendered; he did not make up any samples of mixtures; in rendering the pure lard for standard samples he cut the fat very fine and put it into a large porcelain dish, and adding water, boiled it from forty-five minutes to an hour; then he squeezed out a portion of the fat, and subjected the residue to a little greater heat and extracted all the fat he could possibly get out by any ordinary squeezing method; the lard then contained some water, which he removed as far as he could by decanting; then heated the lard over a sand-bath, being careful not to heat it so much as to burn it; but he certainly had it at a sufficiently high temperature to hold the greater part of the stearine in a molten state, and pass it through the filter.

It would probably depend a little on circumstances which of two samples of lard, one rendered at a low pressure and the other at a high pressure, would contain the most stearine; speaking casually he should say the one rendered at a high pressure would contain the most; he means by high or low pressure, a greater or less pressure in squeezing out the lard as is ordinarily done in a small way; he should think there might be a tolerable variation in the quantity of lard stearine from this cause, but of the pure stearine there ought not to be such a remarkable difference. It depends entirely upon when the lard is produced, when the hog is raised, when killed, what fed upon, and perhaps other conditions, as to how much stearine there may be in lard; authorities differ on the subject; some claim there is as high as 33 per cent., others less, of lard stearine; as far as he has been able to learn, lard stearine varies from 30 to 40 per cent., and pure or chemically pure stearine from less than 1 per cent to about 3 or 3½ per cent. depending upon the conditions he has referred to; his personal examinations have shown a variation of from a little under 1 per cent. up to about 2½ per cent. He should think, the better the hog is, the better would be the

---

* Op. cit., p. 157. †Op. cit., p. 158.

fat from it, and probably the richer the fat in stearine, and this would apply to the lard rendered from a large number of hogs of average fine quality, as compared with the lard rendered from a large number of average inferior quality; samples drawn from different parts of a large tank holding 250 tierces might vary a little, depending upon whether the lard put into it was thoroughly mixed, at what temperature or how fast it had been cooled, and other conditions.

TESTIMONY OF C. GILBERT WHEELER.*

In treating the lard the particular process upon which he relies—although he used others, some of which pointed to the same conclusions and others to no special result—is based upon the insolubility of pure stearic acid in a mixture of absolute alcohol and ether. In treating by that method he takes a given amount of lard and nine times as much of a mixture, composed of equal parts of absolute alcohol and ether, places them in a closed vessel, with a graduated scale upon it, agitates, and exposes to a low temperature, the agitation being repeated a few times; then, after standing for about twelve hours, the supernatant liquid is poured off and as much more of a fresh supply is added, and again shaken; after standing again for about twelve hours the liquid is entirely poured off, the residue collected, dried and weighed, and its amount compared with that obtained in the same way from both pure lard and impure lard. Pure lard should give a certain per cent. of residue, impure lard gives more; the residue so obtained is pure stearine. In the case of pure lard rendered by himself for a standard of comparison in his investigations of the samples in question, the amount of residue obtained was nine-tenths of 1 per cent. In the samples he was to examine, No. 1 gave 3.6 per cent., No. 2 gave 3.29 per cent., No. 3, gave 2.75 per cent. The process was conducted at a temperature of about 75 degrees Fahrenheit; the amounts of lard taken were, for the pure lard, 10 grams; of sample No. 1, 10 grams; No. 2, 5 grams, and No. 3, 5 grams.

The evidence of which an abstract has been given was on the side of the prosecution and the charge of adulteration appears to be well founded in the light of the evidence given. Following is a brief abstract of the chemical evidence introduced by the defense:

TESTIMONY OF DR. ROBERT TILLEY.†

Dr. Tilley made a microscopical examination of the samples *in lite* and gave the following certificate:

CHICAGO, *June* 29, 1883.

This is to certify that on the 9th day of June, 1883, I received from Prof. W. S. Haines samples of lard labeled, respectively, No. 1 Fowler, No. 2 Fowler, No. 3 Fowler; that I have examined the same microscopically, and that I can find no evidence of adulteration; and consequently, in the absence of such evidence, I believe said samples to be pure lard.

ROBERT TILLEY, M. D.

Dr. Tilley added the following explanatory remarks: ‡

He has examined fats when crystallized in the manner described by Mr. Hoskins: he has never had a sample of pure palmitine to examine, and therefore can not say whether there is any difference between the crystals of it and the crystals of stearine; he can not say whether the appearance of the crystals, in the methods used by himself, depends upon the relative proportions of stearine and palmitine, because he does not know of any means, apart from temperature and pressure, of thoroughly isolating palmitine and he has never done it; he is of the opinion that the greater part of the solid fat of either the beef or the hog is stearine, and not palmitine; the constit-

---

* Op. cit., p. 160. † Op. cit., pp. 165, 166. ‡ Op. cit., pp. 167, 168.

nent parts of lard are oleine, palmitine and stearine: the crystals of stearine seem to be modified according as the substance is mixed with beef stearine, or hog stearine; his answers are in part from the books and in part from his own experience: he has never had the opportunity of observing any modifications in the crystals by an increase or diminution of the palmitine in a specimen, and consequently can not say whether or not the crystals would be modified as stearine or palmitine predominated. In applying the method of microscopical examination to lard, described by Mr. Hoskins, he takes a microscopical slide and cleans it, chemically, then puts on the slide a small quantity of the specimen to be examined, puts it in a water-bath, at the temperature of boiling water, covers it with a covering glass and allows it to cool as slowly as possible; he has made a number of experiments of this kind, sufficient to satisfy him that he could find no distinction. The method he used, and upon which his conclusions as to the lard in question in this case are based, was as follows: He dissolved the specimens in sulphuric ether, allowed the ether to partially and slowly evaporate and crystals to deposit, then decanted the remainder of the ether, washed with ether and decanted again, then treated with absolute alcohol; after which the crystals were examined under the microscope.

* * * * * *

Commercial stearine may or may not include palmitine, according as heat and pressure are used in the separation; he thinks he understands the manufacture of commercial stearine; it would depend upon the temperature to which it is subjected whether it would contain all the original parts of the fat, except what oleine may have been pressed out; palmitine is said to liquify at a temperature of 45° C. consequently if the fat material is subjected, at that temperature, to a pressure similar to that used for extracting the oleine, the palmitine would also be forced out; he does not know, as commercial stearine is usually made, that it is ever subjected to a temperature high enough to press out the palmitine, but he has seen it subjected to a temperature in the manufacture of candles—obtaining stearic acid—when it certainly would be pressed out; as to whether the crystals produced from commercial stearine would be the same as those produced from what has been called by other witnesses chemical stearine, he can only say that he is not acquainted with the peculiar characteristic features of palmitic crystals: the size and general character of the crystals depends on the temperature and slowness of evaporation; there is a difference between the crystals of pure stearine from lard and that from tallow; he states this upon the theory that the stearine he has obtained was pure stearine, but inasmuch as he is not sure that the stearine he obtained did not contain a mixture of palmitine, he desires to make that qualification. As between the crystals obtained by his process and those that caused the so called grain of the lard, he could not, chemically, see any distinction whatever; microscopically this grain seems to be the crystal, in a form very much resembling an ordinary roadside bur, when examined under the microscope, but the details are so ill-defined that it is simply impossible to make any differentiation; the crystal itself would be a solution, if melted. He does not think crystals obtained from lard itself, without extracting the oleine, would be as valuable as those from the stearine, because it is acknowledged that that there is at least 60 per cent. of oleine in the lard, and that 60 per cent. of extraneous matter would, he thinks, necessarily render the crystals more difficult to differentiate than if crystallized from the stearine alone: probably the crystals which characterize the grain of the lard are the same as those obtained by the method described by Mr. Hoskins; they might be the same thing, but yet so modified by the influence of the oleine, that the peculiarities of the crystals would be less prominent; they would be more or less stunted, or their development favored. As to whether or not the method requiring the least manipulation would be likely to give the best results would depend entirely upon the advantages gained by the manipulation, and the care with which it is done. In transferring the crystals to the microscope, he takes a small glass tube—a pipette—cutting off for each observation, with a file, the piece used in the previous case, so that it is perfectly clean as it goes into the liquid;

it is difficult to get the crystals under the microscope in perfect form; but he thinks the Board will have an opportunity to see how perfectly they can be gotten out, by an exhibition of photographs of those they have used for this examination; it is necessary that the crystals be washed in order to obtain a plain, clear-cut specimen.

CERTIFICATE OF MICROSCOPIC EXAMINATION BY DR. W. T. BELFIELD.[*]

CHICAGO, *June 29, 1883.*

On June 12, 1883, I received from Prof. W. S. Haines three samples marked "Fowler 1," "Fowler 2," "Fowler 3," respectively. I have submitted these samples to microscopical examination for the purpose of detecting the presence of beef tallow. By either one of two methods I have satisfied myself of my ability to detect the presence of beef tallow in lard, whenever the admixture contains 10 per cent. or more, by weight, of the tallow.

By neither of these methods did I detect the presence of tallow in the samples above mentioned; I am therefore convinced that these samples do not contain an amount of tallow equal to 10 per cent. of the weight.

I have as yet no knowledge of any methods of microscopical examination whereby I can detect an admixture of less than 10 per cent. of tallow *with certainty*, but I have never obtained in the samples above mentioned any appearances other than those which may be presented by pure lard.

WILLIAM T. BELFIELD.

Dr. Belfield further says:[†]

He is not familiar with the manner of manufacturing prime steam lard; he supposes the tanks are covered during the process of rendering and that when put into tierces it is covered. If he found in a sample of lard that had been kept covered an excess of cotton fibers of special characteristics he should not draw any inference as to how they came to be there; if lard in which cotton fibers were found was chemically tested for the determination of the presence of cotton seed oil, and the chemical test was supposed to detect it, he would not be inclined to attach any more value to the chemical test on account of finding individual cotton fibers in the lard by a microscopical examination. He does not wish to be understood as saying no one can detect cottonseed oil in lard by microscopical examination; he meant to say he could not do it. He has not examined the crystals from isolated stearine or isolated palmitine. In crystallizing stearine and palmitine, in the manner described by Mr. Hoskins, he should think the appearance of the crystals would depend on the relative proportions of each, but as he has never worked isolated palmitine he can not say certainly. In the method he uses the appearance of the crystals does not depend on the relative proportions of the stearine and palmitine in the mixture on which he operates; there is said to be a difference between the crystals of stearine and palmitine. He ordinarily transfers crystals from the liquid to the microscopic slide by means of a clean pipette; it can be done, and he has done it, by means of a clean knife blade, or something of that sort. The objection to Mr. Hoskins's method is that the characteristic crystals of lard and tallow are not formed by it; there is so much granular matter that it gives crystals so nearly alike that he can not differentiate between those of lard and of tallow; a principle to be worth anything, in matters of this kind, must detect minute adulterations. The question of discovering adulteration in lard by microscopical examinations is a new one, at least it is so to him. New discoveries of facts by microscopy, which may be subsequently well established, are sometimes questioned even by experienced microscopists, if they fail to follow the methods and directions of the discoverer, as has been the case in some notable instances. These persons may attempt to follow the discoverer, but, doing so imperfectly, fail to secure the expected results. This fact, however, does not apply to his judgment on the method pursued by Mr. Hoskins, because he followed Mr. Hoskins's method as he described it.

[*] Op. cit., p. 176.   [†] Op. cit., p. 177.

CERTIFICATE OF PROF. PLYMMON S. HAYES.[*]

CHICAGO, *June* 29, 1883.

On the 9th day of June, 1883, I received three samples of lard from Prof. W. S. Haines, marked, respectively "No. 1 Fowler," "No. 2 Fowler," and "No. 3 Fowler." These samples of lard I examined microscopically, after having crystallized the stearine found in the samples from solution, and was not able to detect any beef stearine whatever.

PLYMMON S. HAYES.

Professor Hayes adds the following observations: [†]

He has examined crystals which he obtained by dissolving the stearine of beef tallow, of mutton tallow, and of lard—three samples in absolute alcohol; the substances were dissolved and crystallized, redissolved—and recrystallized, and finally the alcohol was evaporated off; he considered those the crystals of pure stearine; he has never examined the crystals of pure palmitine to his knowledge; he does not know what, if any, difference there may be between the crystals from pure stearine and pure palmitine; he does not think the appearance of the crystals obtained by himself depend entirely upon the relative proportions of stearine and palmitin in the specimens; the crystals from a fat, consisting of oleine, stearine, and palmitine, which has been warmed and then slowly cooled, would be those of stearine and palmitine, probably more or less modified, as one or the other was in excess, and also by the presence of the oleine. Molten lard is a solution of oleine, stearine, and palmitine. When this is allowed to cool slowly and crystallize the crystals are valueless, for the reason that they give no distinctive difference when examined plainly or by means of the polariscope. He has examined crystals in the manner described by Mr. Hoskins; he will not say Mr. Hoskins's method is wrong, either in principle or in application, but it does not produce results that are plainly marked. He pursued that process for over two weeks before he abandoned it as useless. In his examinations by that method he took the sample to be examined and put it on a microscopic slide, put a cover-glass over it, and applied heat until it was thoroughly melted, and then set it aside to cool; sometimes he allowed it to cool very slowly, and sometimes he cooled it by means of a cooling apparatus; in neither case did he get satisfactory crystals; he never tried cooling it by means of a hot iron allowed to cool slowly with the lard.

CERTIFICATE OF DR. I. N. DANFORTH.[‡]

CHICAGO, *June* 29, 1883.

I hereby certify that on the 8th day of June, 1883, I received from the hands of Prof. Walter S. Haines, of Chicago, three specimens of lard, numbered 1, 2, and 3, respectively, and said to have been manufactured by the "Anglo-American Packing and Provision Company," of Chicago; that I was requested by said Haines to examine said specimens microscopically, and state my opinion as to their purity or impurity; that I have examined said specimens as thoroughly and carefully as the time allowed would permit, and am of opinion that they are composed entirely of the fat of the hog.

ISAAC N. DANFORTH.

Dr. Danforth in explanation said: [§]

He has examined the crystals of pure stearine; he procured the first specimen from Professor Haines, and afterwards prepared specimens by methods used by chemists; he has not examined the crystals of pure palmitine; he does not understand that any

[*] Op. cit., p. 179.
[†] Op. cit., pp. 179, 180.
[‡] Op. cit., p. 182.
[§] Op. cit., p. 184.

method has been devised for absolutely isolating palmitine, and he does not know whether or not there is any difference between the crystals of pure stearine and those of pure palmitine; he can not say whether the crystals procured by him depend on the relative proportions of stearine and palmitine in the substance from which they are procured; while he does not claim to speak as authority on chemical subjects, he thinks the substance from which these crystals were obtained was particularly pure stearine; he can not positively say whether the crystals obtained from a fat consisting of oleine, stearine, and palmitine would be those of stearine and palmitine more or less modified as one or the other was in excess; he thinks the presence of the oleine would modify the crystals somewhat, but to what extent he is not able to say.

METHODS FOLLOWED IN MAKING MICROSCOPIC INVESTIGATION BY MESSRS. TILLEY, BELFIELD, DANFORTH, HAYES, AND ROSE.[*]

Dr. Tilley's formula.—For the production of the crystals of stearine, whether from beef or hog products, I dissolve the fat in question in sulphuric ether, 10 grains in 2 drachms; allow the ether to partially and slowly evaporate, and consequently crystals to deposit; decant the ether remaining, wash with ether, decant again and treat with absolute alcohol, then examine the crystals under microscope.—Robert Tilley, M. D.

Dr. Belfield's formula.—Ten grains of the fat are dissolved in Squibb's ether; the quantity of the latter may be 1 drachm or 2 drachms, or instead of the ether, a mixture of this substance with absolute alcohol in equal parts may be employed as a solvent. The solution is allowed to stand in a test tube uncorked for twenty-four hours, at ordinary temperature; at the expiration of this time crystals are observed at the bottom of the tube; these may be examined directly through the microscope, or supernatant liquid can be poured off and replaced by a drachm of absolute alcohol; this is subsequently removed and the crystals examined; the crystals may be mounted for examination in the solvent used or prepared dry. When specimens of pure lard, pure tallow, and mixture of both, within certain proportions, are treated in this way, characteristic crystals are formed by means of which the identity of the specimen can be established. Essentially the same results are obtained when the method is varied by changing the quantity of the solvent, within certain limits, or by repeated washings with alcohol.—William T. Belfield.

Dr. Danforth's formula.—The oleine is first extracted from the specimen to be examined by the use of absolute alcohol or a mixture of alcohol and ether in equal parts, or by any other method the experimenter may choose to adopt; the remaining stearine is then dissolved in any one of a number of menstrua, as, for example, ether, turpentine, benzole, oil of Scotch pine, or any other solvent of stearine; I usually employ turpentine; from this solution crystals are allowed to form, and the resulting crystals are then mounted upon glass slips prepared for the purpose in Canada balsam diluted with twenty-five per cent. chloroform, or a solution of damar, or they may be examined in the original solvent. Another method I have used recently to a considerable extent, is to place the specimen to be examined immediately in some solvent of stearine, as ether, or benzole, or turpentine (usually the latter), in the proportion of ten grains of the specimen to be examined to a fluid drachm of the solvent, without first extracting the oleine: I then watch carefully for the formation of the first crop of crystals; these crystals are then immediately examined after being mounted in either one of the media that I have already mentioned. Recently I have employed a solution of balsam in chloroform as a mounting medium because it gives the clearest field. The crystals are examined by the use of a one-quarter inch objective and an "A" eye-piece, giving a magnifying power of about two hundred and fifty to two hundred and sixty diameters. The difference in appearance between the lard stearine and beef stearine crystals is clear and definite.—Isaac N. Danforth.

---

[*] Op. cit., pp. 197, 200, 201.

Professor Hayes's formula.—Take a suitable vessel that can be well corked and put in it some of Squibb's ether, add to it lard, or preferably the stearine from the lard in question, until at a temperature of 85° F. A slight portion of the lard or stearine remains undissolved (the stearine may be obtained from lard by making a hot saturated solution of lard in a mixture of equal parts of absolute alcohol and ether, or some other solvent, and crystallize out the stearine by cooling the solution): the vessel is then very tightly corked and the ether heated until the solution is perfectly clear; then the vessel is placed in water, or left in a room, at a temperature of from 60° to 70° F.: the slower the solution cools the larger will be the crystals: the ether is then decanted from the crystals and the crystals washed with cold ether or a mixture of absolute alcohol and ether; the crystals are again washed with absolute alcohol, after which a fresh portion of absolute alcohol is put on the crystals and allowed to stand. The crystals may be mounted either in Canada balsam, glycerine jelly, or dry. In examining the crystals I usually employ a fifth objective and a "B" eye-piece. The difference in form of the beef and lard stearine crystals furnishes a ready means for their detection.—Plymmon S. Hayes.

Dr. Rose's formula.—Take 1 gram of the fat and dissolve in absolute alcohol, allow it to stand until crystals form, decant off the solution, and wash with repeated portions of absolute alcohol, finally remove the alcohol by heat, and dissolve in turpe..tine and set aside to crystallize; when crystals form remove with a pipette and examine them with the aid of a microscope; the crystals formed from beef products are quite different from those obtained from hog products.—P. B. Rose.

AN OUTLINE OF DR. ROSE'S METHOD OF CHEMICAL ANALYSIS.*

Take a definite quantity of the article to be examined, add nine times its weight of a mixture of equal parts of ether and absolute alcohol (the absolute alcohol used was but 93 per cent. alcohol), thoroughly agitate, place the jar in water, kept at 65° F. for twenty-four hours; the clear portion is then decanted off and about one-half as much more of the alcohol and ether is again added; agitate and allow the jar to stand for twenty-four hours in water at the constant temperature of 65° F.: the liquid portion should then be decanted off and finally filtered; the filter containing the residue placed in a jacketed filter filled with water and kept at the temperature of 65° F.; the residue on the filter is washed with a mixture of alcohol and ether, cooled to the temperature of 65° F.; the filter and residue are dried and weighed, the filter being counterpoised by a second filter.

The following are the amounts taken and the results:

	Grams.
Prime steam lard taken	10.905
Residue obtained	2.846
Kettle-rendered leaf lard taken	10.351
Residue obtained	3.259
No. 1 sample taken	10.065
Residue obtained	2.550
No. 2 sample taken	9.979
Residue obtained	2.524
No. 3 sample taken	9.599
Residue obtained	2.409

The other experiments were conducted in like manner except that the temperature was kept at 50 and 66° F., respectively.

P. B. Rose.

*Op. cit., p. 202.

CERTIFICATE OF WALTER S. HAINES.*

CHICAGO, *July* 2, 1883.

This is to certify that I have examined the specimens of lard delivered to me by Col. Ezra Taylor and numbered 1, 2, and 3, and also the specimens placed in my hands by Dr. Robert Tilley, numbered 1, 2, and 3, and that I find them pure lard and unadulterated with tallow or cottonseed oil.

WALTER S. HAINES.

METHOD EMPLOYED BY PROFESSOR HAINES.†

To test for the presence of cottonseed oil I used the color test with strong sulphuric acid: from 5 to 10 grains of the lard are stirred up with one or two drops of the acid and the color produced is noted. Pure lard thus treated gives a color ranging from salmon to slate, cottonseed oil a dark olive brown, while lard mixed with cottonseed oil produces a well-marked mixture of the two colors.

To test for the presence of tallow I make a solution of lard in some slightly warmed solvent, such as turpentine, ether, or ether and alcohol; after a number of hours the crystals thrown down are examined (preferably after washing with a little absolute alcohol, or ether and alcohol) by the aid of a microscope; the form of crystals indicates whether the lard has been adulterated or not.

WALTER S. HAINES.

In explanation of his results, Professor Haines said:‡

In saying that he used essentially Professor Delafontaine's process, he meant that he had endeavored to follow the spirit of that process as he understood it. Professor Delafontaine insists that his examinations of the lard were all conducted comparatively with samples of known purity; and as his own were conducted in that way, it makes but little difference what temperature is employed; he always used the same temperature for the examination of the samples in question. He thinks Professor Delafontaine's process is not sufficiently accurately given to admit of being followed with entire precision; he does not now remember the temperature Professor Delafontaine used; he read that gentleman's certificate and evidence in this case in respect to his process. He (Prof. D.) states that any chemist, on reading his certificate, will be able to tell his process. The certificate does not allude to temperature, and he concludes that, as nothing was said in it about the temperature employed, it was considered by Professor Delafontaine as of minor importance, or no importance; it is true, on cross-examination, Professor Delafontaine did mention the temperature, but he concludes that when a chemist presents with deliberation, in writing, an account of his process, and states, in connection with the certificate, that it is amply sufficient for any chemist to follow, and has excluded any mention of temperature in the certificate, the party attaches to temperature a very trivial importance, or no importance at all; it might have been considered an accidental omission, but for the fact that it was insisted that he should give all essential particulars of his process, and after the certificate was submitted, he, again and again, insisted that everything essential was stated in the certificate; and in view of these facts he (the witness) can not consider that the omission was by an oversight, especially as when, on the cross-examination, the temperature was developed it was not then spoken of as being essential. In all his experiments of treating the lard under consideration, concerning which he has testified, he either did the work himself, or saw it done from beginning to end; those performed in his own laboratory were conducted entirely by himself, in person, and whenever he left the laboratory, while the process was going on, the room was securely locked.

---

* P. 219.   † Op. cit., pp. 219, 220.   ‡ Op. cit., pp. 220, 223.

In a mixture of 90 per cent. of lard and 10 per cent. of cottonseed oil, treated with sulphuric acid, as he has described, the color produced will be the salmon, with a tinge of slate peculiar to the lard, and in addition a tinge of the olive brown peculiar to the cottonseed oil; the shades of color produced by this process will probably be somewhat differently classed by different individuals, even as a result of seeing identical colors or shades of color; some might call what he describes as olive brown, a mahogany brown, and so of other tinges of color, depending upon peculiarities of vision in different individuals; some persons are slightly color-blind, others greatly so; he speaks of these colors as they appear to his own vision; if a person has not an eye trained to distinguishing colors, he might not, perhaps, discover the difference in these shades; the gentlemen who have been associated with him in these examinations have been unanimous in picking out the colors peculiar to the lard, and to the cottonseed oil, and became trained in that respect before pronouncing on the several samples; he has never applied the test to crude cottonseed oil, but he has experimented on a number of specimens of commercial oil, and refined oil and cottonseed stearine; his observation has been that the refined oil does not give as marked a color as the commercial oil does; he can not say whether or not there would be any difference in the oil made in summer or winter; he has concluded that the color comes from the oil itself, and not from any foreign substances that might be in it, because he has tested three specimens of fine oil, which were entirely colorless, and has also tested perfectly colorless cottonseed stearine, and from these tests he concludes the color produced by the test is due to the oil itself. In respect to the variation of the amount of stearine in different samples of lard, he accepts as evidence of its truth the agreement of authorities who have discussed it in books and other publications; he understands the agreement of writers, on this point, to rest on the well-known fact that in lard are the constituents of stearine and palmitine, which are solid oils; these can be separated approximately, not perfectly, perhaps, but sufficiently nearly so, when treated alike, to enable one to determine that some lards contain more stearine than others.

* * * * * * *

His attention was first directed to the subject of the detection of tallow in lard about five or six years ago; at that time he made some chemical experiments on the question, and again, about eighteen months ago, he made a few other experiments in the same direction, and within the past four weeks he has made numerous experiments. Husson's method, as published, is to take a mixture of alcohol, at 90 degrees, and ether at 66 degrees, which he understands to mean 90 per cent. alcohol and absolute ether; with this mixture he treats the previously warmed fat and allows the more solid portion of it to crystallize; he has used Husson's method, as he read it in French, and failed to obtain satisfactory results by it, and considered it untrustworthy; he does not fully condemn it, because he is not fully convinced as to what Husson means by alcohol at 90 degrees and ether at 60 degrees; if his interpretation of the method is correct, he does not concur in its being of value. The comparative results obtained by Professor Delafontaine and others, who have testified on behalf of the prosecution in this case, by which the lard in question was shown to have more stearine than the pure lard, which they used in comparison, carries no conclusion to his mind whatever, for the reason that Professor Delafontaine testified that the only lard which he knew to be pure, and with which he tested the samples in comparison, was kettle-rendered lard. It is manifestly unfair to take that as a standard for comparison with prime steam lard; but besides that, his own experience in treating lard, by the process described by Professor Delafontaine, shows it to be fallacious from the very foundation, and he attaches no importance to any results obtained by it or by modifications of it.

He means to be understood that, so far as he is able to determine, the samples of the lard now under consideration, which were examined by him, contain absolutely no adulteration.

CERTIFICATE OF PROF. E. B. STEWART.*

JULY 3, 1883.

This certifies that I have carefully examined three specimens of lard received from Prof. W. S. Haines on the 20th of June, 1883, and find that they present the characteristic of pure hog's lard, free from tallow and cottonseed oil.

E. B. STEWART.

METHOD OF EXAMINATION FOLLOWED BY PROFESSOR STEWART.†

Both pure kettle-rendered and steam-rendered lard were treated with about three times its weight of absolute alcohol at a temperature just sufficient to melt; the solid residuum which separated on cooling was assumed to consist of tristearate of glyceryl principally; this was treated with, first, oil of turpentine; second, petroleum naphtha; third, bisulphide of carbon; fourth, benzole; fifth, Squibb's strongest ether; and, sixth, melted in balsam fir. Pure beef tallow was treated with absolute alcohol in the same way, and subsequently with the same reagents.

E. B. STEWART.

JULY 3, 1883.

CERTIFICATE OF PROF. S. P. SHARPLESS.‡

JULY 3, 1883.

I have examined three samples of lard submitted to me by Prof. W. S. Haines, and marked Nos. 1, 2, 3, Fowler Bros. I have been unable to find any adulteration in these samples, and believe them to be pure hog product.

S. P. SHARPLESS.

In explanation of his results Professor Sharpless says: §

He received from Professor Haines, two weeks ago, three samples of lard, marked, respectively, No. 1, No. 2, and No. 3, since which date he has devoted his time to their examination. The samples were received in tin boxes, wrapped in paper, and properly sealed. The work of examination was commenced on the 19th of June, and was conducted jointly by Professor Doremus, Professor Haines, and himself, at the laboratory of the College of the City of New York. After opening the boxes the contents of each were thoroughly mixed, and then 5 grams were weighed out from each of the boxes; these specimens were submitted to the action of absolute alcohol and the strongest ether, both being carefully tested as to their strength; he, at the same time, for the purpose of comparison, prepared a sample of tallow, rendering it himself in order to be certain of its purity; 5 grams of this tallow were also weighed out; these samples were weighed into small flasks, and the fats of each were melted, and then 50cc of the mixture of alcohol and ether was poured over each specimen; the flasks were shaken until the alcohol and ether had completely dissolved the whole of the fat. This process is a test that will show whether the lard contained starch or salt or whether there is much water in it; lard having much water in it will give a clear solution, but will be milky in appearance; these lards all gave perfectly clear solutions, with perhaps an occasional particle of wood from the cask; there was very little fibrous matter in any of the samples; he has never yet seen samples of lard that were perfectly free from fiber; these were as free from it as any he had ever seen; the flasks were set in a closet, at the same temperature for each, until the next morning, when they were examined. The bulk of the precipitate in the different flasks differed; in some it was slightly flocculent, in others it adhered to the bottom of the flask; this latter condition was more marked in the case of this tallow, which formed a thin layer over the bottom of the flask; the liquid was poured off and 25cc of fresh was added, and the flasks were allowed to stand, with the fresh solvent in them, until the next morning, and

---

* Op. cit., p. 228. ‡ Op. cit., p. 233.
† Op. cit., p. 228. § Op. cit., pp. 229-230.

then their contents were filtered, the precipitate washed with a little absolute alcohol, and weighed. Sample No. 1 gave 4.35 per cent. of precipitate; No. 2 gave 2.99; No. 3 gave 2.4, and the tallow gave 5.3 per cent. of precipitate. At the same time another series of experiments was tried by weighing out the same amount of lard and adding to it the mixture of alcohol and ether; in this case the lard was not melted previous to the addition of the alcohol and ether; after these were added the flasks were shaken thoroughly and stood in a water-bath at the temperature of Croton water, which at that time was about 72° F., the water being allowed to run around the flasks; they were left standing in the water in this way for a little over two hours, during which time they were shaken every fifteen minutes; at the end of this time the contents were allowed to settle; the liquid was then poured off and 25cc more of the solvent was added, again shaken, and the residue of each sample was collected on filters, which had previously been weighed; this series of tests produced a little higher per cent. of residue in every case than the former. No. 1 gave 5.63; No. 2 gave 4.17; No. 3 gave 3.69, and it was found impossible to filter the pure tallow on account of the large mass of crystals it gave. The day after this test they received a sample of pure prime steam lard from Chicago, rendered under the supervision of Dr. Tilley; he made two parallel experiments upon this pure lard; the first one gave a residuum of .44 of 1 per cent.; the second gave 3.14 per cent. These samples were taken from the same mass of lard at the same time, and were treated with exactly the same solvent of alcohol and ether; during the process they stood side by side in the same closet, were filtered off at the same time, weighed at the same time, and in every way treated alike, and yet one gave nearly eight times as much residue as the other. In connection with Professors Remsen and Witthaus he tried another series of experiments, following out the same method as in the first series, with results substantially the same as in the first experiment, which he has detailed so far as the lard was concerned. In this experiment there was also a mixture of lard and tallow, but he did not find that the addition of the tallow made any perceptible difference in the result. This experiment will probably be more fully described by Professor Remsen

CERTIFICATE OF PROF. IRA REMSEN.*

CHICAGO, *July* 5, 1883.

I hereby certify that I have examined by chemical methods the three samples of lard designated as Nos. 1, 2, and 3, "Fowler," submitted to me by Prof. W. S. Haines, and have failed to find any foreign substance in them. I am, therefore, of the opinion that the samples are pure lard.

IRA REMSEN.

Professor Remsen† stated that he received from Prof. W. S. Haines, for analysis, three samples of lard, designated 1, 2, and 3, "Fowler," and has been working on them constantly for about ten days, to a considerable extent night and day. When these samples were first submitted to him he set about a very careful search through the literature on the subject to determine what method ought to be adopted in the examinations of them. He was disappointed by finding that the chemical study of lard had, apparently, received very little attention. The methods for the chemical examination of lard which have been, perhaps, the most frequently employed, are similar to those which are used in the examination of butter; indeed, the chemical knowledge of butter is much more general than that of lard. After considering the subject he decided, as he thinks most do who are called upon to investigate lard, to adopt as an experimental method that of Husson, which is based on a very simple principle. Fats are known to differ in the proportions of oleine, palmitine, and stearine contained in them. In liquid fats there is a larger proportion of oleine, and a less proportion of stearine than in those of a more solid character. In case a fat

---

* Op. cit., p. 246.   † Op. cit., pp. 238, 239, 240, 241, 245, 246.

which is naturally rich in stearine should be adulterated with one which is rich in oleine, he thinks chemistry could very easily detect that adulteration; there are, however, several fats which are so similar to each other, in respect to the proportions of the constituents they contain, that when they are mixed it is a very difficult matter for chemistry to detect the mixture. The best method he could think of was the separation of the stearine from the oleine, if that could be effected; but there is no method by which this can be entirely or anything like entirely done. If fats are treated, in comparison, with some substance which will partly separate the stearine, but leave some behind, and then that which is left behind in the one is compared with that left behind in the other, both being accurately ascertained, and one is found to be much larger than it should be, it is strong ground for suspicion in one's mind that there is something the matter with it, without, perhaps, being able to say exactly what has been put into it. In order to test a sample of lard by this process it must first be known what standard lard really contains, or, if tallow or any other fat, what that fat contains; this, however, can not be ascertained from the books, because, in respect to reliable data on the subject, they are singularly silent. It is stated repeatedly, in the books, that lard varies in its composition, depending on a variety of causes, such as the dryness and other conditions of the food on which the animal from which it was produced was fed, the season of the year when fed or when killed, etc. No one seems to have made so complete an investigation of the subject as to state to what extent these variations may go, hence it is absolutely necessary, in an examination of lard, to first get something that may be considered a standard of pure lard, to know what is pure lard, and to make an exhaustive investigation of the subject. There ought to be a very extensive investigation of different specimens of lard, so as to find out what variations in the constituents are possible.

For the purposes of this investigation, he procured at the outset a sample of lard known to be pure, with which comparisons could be made as the investigations progressed; and then if, on a comparative examination, the lards submitted for examination were found to conduct themselves in all respects in the same manner, and no differences were found in them, the conclusion would be justified that the lards to be tested were pure. In the process of examination, he in the first place applied Husson's method, and also a modification of that method, suggested from reading the testimony of Professor Delafontaine. He also applied the elaidine test and the pattern test, and he treated them with sulphuric acid. He also examined these samples by means of the spectroscope and by transmitted light, and he has to some extent examined them microscopically; and he can say that, after all the examinations which he has been able to give to the samples of lard submitted to him by Professor Haines, his only conclusion is that he can find no impurities whatever in them.

In his examinations he has paid particular attention to the method described by Professor Delafontaine, in that gentleman's testimony in this case. That is not the method described in the books, and known as Husson's method, but is a modification of that method to suit the case of lard. It depends upon the relative amount of residue remaining after treatment with alcohol and ether, and he has to say that if that method is a good and reliable method, then, beyond any possible question in his mind, the samples 1, 2, and 3 submitted to him are pure lard. If it is not a reliable method it proves nothing. He is not prepared to absolutely condemn the method, for the subject has never been studied, so far as he knows, with that care that would warrant the basing of positive conclusions upon that process of determining it. It is possible that there is the germ of a good method in it; but, as described by Professor Delafontaine, he is quite confident it could never be used to prove positively whether lard is pure or impure. In respect to Husson's method, without any modification, he should say most emphatically that it is not a reliable process for determining the adulteration of lard. The Blythe pattern process is, as he understands it, the same as what is known as "cohesion figures." There is some confusion in the use of terms in describing matters of this sort. He is, however, quite sure that the process called the

"Blythe pattern process" and "cohesion figures" is one and the same thing. Of this process Mr. William L. Carpenter, in the "Journal of the Society of Chemical Industry," of London, under date of March 29, 1883, says:

"In reply to Mr. Newland's inquiry on this subject, I may say that when Professor Tomlinson first brought them forward I spent several weeks in fruitless endeavors to apply the method to analytical examination of oil, and the result, I regret to say, was a complete failure."

He has read the testimony of Mr. Hoskins given in this case in respect to his chemical analysis of lard. Mr. Hoskins washes out a little differently from Professor Delafontaine, but he regards Mr. Hoskin's process very much as he does that of Professor Delafontaine. He has also read the testimony of Mr. Hirsh, and will say that he considers the chemical process pursued by Mr. Hirsh as the least reliable of any of those referred to. Mr. Hirsh made use of the same principle that the other two chemists named did, or attempted to do so. Mr. Hirsh, taking the same quantity of the lard to be examined and of pure lard, and applying his process, seeks to compare the residue obtained in each specimen; that principle is the basis of the methods of all these gentlemen; the others weigh the residue obtained, which is the only possible way to deal with it chemically, but Mr. Hirsh measures it; he has never before heard of measuring a precipitate; that is something entirely novel and original.

In prosecuting this investigation he tried the color test for cottonseed oil, with sulphuric acid; this consists of taking a known quantity of the specimen to be examined and dropping on it two drops of sulphuric acid; with cotton-seed oil the effect of this combination is to produce a change of color. This method was tried with cottonseed oil, with mixtures of cottonseed oil and lard, one of which was as low as 10 per cent. of the oil, and with the samples 1, 2, and 3 (Fowler lard); in the mixture of lard with 10 per cent. cotton-seed oil, he could positively identify the presence of the cottonseed oil, but he could not detect any evidence of it in the samples 1, 2, and 3. This test depends for its success upon having the right conditions; it is an extremely delicate test, and must be made under certain conditions in order to get any results at all. He has since tried the same process with another sulphuric acid and failed in being able to distinguish one from the other: that experiment did not prove anything to his mind.

The subject of investigations for the detection of cottonseed oil or tallow in lard is one of the most complicated with which chemists have to deal. When chemists say they can not solve such questions people are apt to laugh at them. Altogether too much is expected of chemistry in some cases, and in others not enough of credit is given. The subject of investigation of fats has been worked on for many years, and all the methods which have been employed have been, in general, found unsatisfactory. In the case of butter the question has, within the past ten years, been studied with great care, and in consequence it is now possible to tell positively what the nature of butter is; other fats have not been examined with the same care, owing to the immense amount of time and labor necessary to go through the investigation fully. In a general way he will say that the methods employed for determining the question are unsatisfactory to him. In the case of the samples 1, 2, and 3, now in question, he can say, that with the investigations he has been able to make of them he has found no evidence whatever of impurity: All the tests he has applied to them, so far as they have given indications, have indicated the absence of impurity; the methods being imperfect, he can not say positively that the lard is pure, but the indications are all in that direction, with no indications whatever in the opposite direction.

* * * * * * * *

The spectroscope develops a difference in the appearance between cottonseed oil, either alone or mixed with lard, and that of pure lard; by the use of the spectroscope he was able to positively, and without difficulty, tell that a certain specimen, unknown to him at the time, but which really contained 10 per cent. of cottonseed oil, had cottonseed oil in it. The samples of Fowler lard did not give the cottonseed oil appearance when tested by the spectroscope, but acted in all respects the same as the pure prime steam lard did. He does not attach much importance to this test, and can only say that so far as the examination of the Fowler lard by this test is concerned, the results were negative.

When examined by transmitted light cottonseed oil has a yellowish color, which neither pure lard nor tallow has, but all mixtures of lard and cottonseed oil have this color. The Fowler lards failed to show any appearance of this color when examined by transmitted light.

The cottonseed oil he used in all these experiments was refined oil. He has never used the bleached or colorless oil for such experiments. The test by transmitted light would be of no value whatever in detecting the colorless oil.

Finally, in respect to the color test for the detection of cottonseed oil: When he testified two days since, he had spoken cautiously and not very confidently of the value of this process. At that time he had just come from the laboratory where they (himself and others) had met with a difficulty which at that time none of them could solve. This arose from their having used something they did not know anything about. This difficulty has, however, been since explained, and he is now fully prepared to make a positive statement in regard to the value of the sulphuric-acid test as a means of detecting an admixture of cottonseed oil in lard, and has joined with Professor Sharpless and others in the statement read by Professor Sharpless in that gentleman's final testimony in regard to the value of that test. The expression of this paper (see pp. 233, 234, which the witness read) is his deliberate judgment upon the question of the reliability of the color test for cottonseed oil; and in view of all his examinations of the samples 1, 2, and 3 of Fowler lard, he is now prepared to express a positive opinion that these samples do not contain any cottonseed oil. He did not, as he has said, use any bleached oil in this process, but other gentlemen did, who have given, or will give, evidence on that point. In all the oil he has used he always gets the color reaction peculiar to cottonseed oil, not always to exactly the same extent, but so sufficiently and clearly marked as to be unmistakable, and the absence of it is proof positive, to his mind, that there is no cottonseed oil in the samples of Fowler lard he has examined.

CERTIFICATE OF PROF. R. A. WITTHAUS.*

CHICAGO, *July* 5, 1883.

This is to certify that I have made chemical examinations of three samples of lard marked, respectively, Nos. 1, 2, and 3, Fowler, without obtaining the slightest evidence of the presence of any impurity. I therefore consider them as being samples of pure prime steam lard.

R. A. WITTHAUS.

Professor Witthaus also exhibited a table showing the results of five experiments with modifications of Husson, showing the varying proportions of insoluble residue left by ether and alcohol. In the first four experiments the residue was washed with 10cc of absolute alcohol, in the fifth with 30cc.†

---

* Op. cit., p. 254.   † Op. cit., table, p. 249.

Substances treated.	Per cent. of residue.				
	No. 1.	No. 2.	No. 3.	No. 4.	No. 5.
Fowler's lard:					
No. 1	5.44	5.06	......	4.65	3.94
No. 2	4.03	3.28	......	3.18	2.88
No. 3	3.64	3.12	......	2.68	1.80
Prime steam lard:					
From the can	3.64	3.27	3.85	3.19	1.69
From the tierce	4.62	3.35	......	......	......
Pure tallow	4.22	3.24	3.79	2.46	......
90 per cent. pure lard, 10 per cent. tallow	......	......	3.42	......	......
80 per cent. pure lard, 20 per cent. tallow	......	......	3.54	......	......
70 per cent. pure lard, 30 per cent. tallow	......	......	3.58	......	......
60 per cent. pure lard, 40 per cent. tallow	......	......	3.28	......	......
90 per cent. pure lard, 10 per cent. cottonseed oil	......	......	......	2.67	......
80 per cent. pure lard { 10 per cent. cottonseed oil / 10 per cent. tallow }	......	......	......	2.36	......
60 per cent. pure lard { 10 per cent. cottonseed oil / 30 per cent. tallow }	......	......	......	2.23	......

The last experiment (No. 5) was tried to ascertain the effect of washing the residue with 30 instead of 10cc of absolute alcohol. The result clearly showed that much depends on the amount of washing to which the residue may be subjected. This residue is not absolutely insoluble in absolute alcohol, and it is probable that by excessive washings it might all disappear.

It does not appear from Professor Delafontaine's testimony that he washed the residue at all; but, inasmuch as some of the liquid containing more or less of the dissolved material, would remain on the filter, unless washed off by the alcohol, and drying, would improperly increase the weight of the precipitate, they deemed the washing with a very small quantity of alcohol necessary in order to arrive at true results. He should not expect to get essentially different comparative results by not washing, and as the whole experiment is comparative it makes little difference whether the washing is done or not.

Professor Delafontaine claims that the presence of tallow is proved by an increase of the residue. The results of the experiments shown in the table prove exactly the contrary, so far as they prove anything in that respect. The admixture of cottonseed oil tends to greatly reduce the amount of the residue obtained by Professor Delafontaine's method.

CERTIFICATE OF W. M. HABERSHAW.*

CHICAGO, *July 5*, 1883.

I have analyzed three samples (sealed) of lard marked Nos. 1, 2, and 3, delivered to me by Dr. W. S. Haines on the 21st of June, 1883, and find them free from adulteration, and, in my opinion, pure lard.

HABERSHAW.

Mr. Habershaw said: †

The analysis of fats has, until the last five years, been a question involving a great deal of doubt. Among the first published methods of treating fats was the color test,

* Op. cit., p. 254.   † Op. cit., pp. 257, 258.

described in a French work by Theodore Chateau. By that test oils and fats were treated with different reagents for producing different colors. He has examined all kinds of oils by that method. In some cases good results are obtained, in others they are unsatisfactory. The method is of somewhat doubtful value. Then came the elaidine test, by which the oleine of a fat is hardened by the action of an oxidizing agent. He does not regard that test as of any value. About 1878, a German method for analyzing oils by means of a standard solution of an alkali was published. He has used that method from then until now, and has found it to produce excellent results. In that method the substance is accurately weighed and treated with hydrate of potash of known value. The result is expressed in milligrams or grams of hydrate of potash required to saponify a stated amount of fat. The standard which is used is 1 gram of fat, equivalent to (blank) milligrams of hydrate of potash. In conjunction with that process, if an oil be examined by the oleate of lead process, which enables one to separate the equivalent oleate of lead from the equivalent stearic and palmitic acid, you ascertain the amount of the stearic and palmitic acids derived by difference, or they may be estimated directly.

* * * * * * *

His work has been entirely independent of other chemists who have examined those samples of lard. The composition of lard is about 47 of oleic acid and 47 of stearic acid; he can not give the chemical formula of lard; the lowest amount of stearine he has found in pure commercial lard was 34 to 40 per cent.; the highest amount was about 45 per cent.; by stearine he means the combination of stearine and palmitine; he has never analyzed lard so as to obtain the tristearine, and has never made the ultimate analysis of either oleine, palmitine or stearine. He does not think there is any difference between the olein of lard and that of tallow; the chemical characteristics of chemically pure stearine are always identical; in the mixture of stearine and palmitine known to the trade as stearine the characteristics would, he supposes, differ, but he can not describe the differences. He believes in the sulphuric-acid test for the detection of cottonseed oil, when used by those who understand it; he has had a great deal of experience with that test and he can detect cottonseed oil by it. He has made a great many analyses of butter; and he has used the Angell and Hehner process, with which he is quite familiar. In the analysis of butter the question of its purity is decided by its percentage of insoluble fatty acids; his own analyses give 87 per cent. as the average of fatty acids in butter; Angell gives 87.34; the range is 2 per cent. either way; if a sample of butter runs over 1 or 2 per cent. above the maximum he has found in pure butter he would condemn it.

TESTIMONY OF PROF. R. OGDEN DOREMUS.[*]

Professor Doremus said he undertook the analysis of the samples 1, 2, and 3, Fowler, and the samples of pure lard by what is called Muter's process, which consists in precipitating the oleic, palmitic, and stearic acids by a salt of lead; this gives the oleate of lead, the palmitate of lead and the stearate of lead; the oleate of lead alone is soluble in ether. After the oleate of lead was removed by this solvent and after filtration, it was decomposed by an acid, and a solution of oleic acid and ether was obtained. A small part of this solution was drawn off, the ether evaporated, and the residue weighed; from this the amount of oleic acid was estimated; the palmitate of lead and the stearate of lead which remained in the filter were removed, decomposed by an acid, and weighed, giving the palmitic and stearic acids combined. Chemistry has not reached that degree of perfection by which these two last-named acids can

---

[*] Op. cit., pp. 251, 252, 253, 254, 255.

be completely separated; therefore they must be estimated in combination; the results of this process were as follows, in percentages:

Acids.	No. 1.	No. 2.	No. 3.	Pure lard.
Oleic acid	60.60	60.42	65.02	65.40
Palmitic and stearic acids	33.85	31.88	30.42	28.16
Total	94.45	92.30	95.44	93.56

The remainder of the substance was glycerine, which being soluble in water was washed away. Lards are liable to variation in the proportion of these acids which they contain, owing to various causes, such as differences in feed of the animals, different seasons of the year in which they are killed, and other causes, and there are also differences in the lard taken from different parts of the same animal. From the fact that the analyses of all these specimens so nearly agree in their proportions of the acids—the variations being only such as are liable to be found in pure lard—he feels justified in stating that he believes the samples, Fowler Nos. 1, 2, and 3, contain the proper proportions of the ingredients of lard and are, therefore, pure.

Analytical chemistry is not capable of determining whether a specimen of stearine is from the fat of the hog or from the fat of the bullock, but these lards, being shown to contain the proper proportions of the constituents of pure prime steam lard, he claims are pure. Muter, as he now remembers it, reports lard as containing a little over 47 per cent. of oleine and about the same percentage of palmitine and stearine, but Muter doubtless referred to lard rendered from the leaf fat alone; in Europe they would not, as we do in this country, designate as lard the fat from all parts of the hog but only such as comes from the leaf.

He has examined under the microscope the specimens of crystals obtained by Dr. Belfield, Professor Hayes, and others, and he believes the microscope is capable of determining the question as to whether the substance from which the crystals were obtained was the stearine of beef or of the hog. He does not claim to be a professional microscopist, but he has used the microscope largely in chemical investigations and in instruction to college students. In examining the crystals of Professor Hayes, slide after slide was placed before him, under the microscope, and without any previous knowledge on his part as to what the specimen was, he was able to at once correctly decide which was from pure lard, which from tallow, and which from mixtures of the two. The difference in the crystals is very marked, and is beautifully illustrated by the photographs exhibited by Dr. Belfield.

In respect to adulteration of lard by cottonseed oil, he believes that gentlemen who are skilled in handling these substances can by the ordinary senses of taste and smell, and by its color, detect an adulteration by it when the adulteration amounts to 5 or 10 per cent., and if the adulteration has been with the common or unbleached oil. he may not, from the color, be able to say that it is certainly cottonseed oil that has been mixed, but he will be able to detect the presence of some abnormal substance. He believes that an expert can, by the sulphuric-acid test, decide whether lard is adulterated with cottonseed oil. There are two methods of doing this. First, dropping the acid upon the lard and allowing it to remain, watching the changes of color. Second, dropping the acid upon the lard and stirring them together, and then watching the development of colors. By this process an adulteration by a substance like cottonseed oil can certainly be detected if it exist to the extent of 10 per cent. He has read the testimony for the defense in this case on the subject of the sulphuric-acid test, and fully agrees with them as to the value of this test, and he has joined in the statement in that regard read by Professor Sharpless.

By the aid of the spectroscope for the examination of lard, cottonseed oil, if present, is at once and clearly indicated; not that it is certainly cottonseed oil, but that there is some foreign substance in the lard. There is nothing in chemical analysis that can compare in delicacy with the spectroscope; the utmost reliance is placed on this instrument, and it is used for the determination of the most serious and delicate questions, and such as involve the issues of life and death. There are two forms of this spectroscope, that of direct vision, and the micro-spectroscope; the latter being a combination of the microscope and the spectroscope; both were employed in the examination of the samples of the lard now in question. A piece of rubber was arranged with small cavities cut into it, and into one of these cavities was placed a specimen of pure lard, which had been previously melted and filtered; in another cottonseed oil; in another lard with an admixture of 10 per cent of cottonseed oil; in another lard mixed with 20 per cent. of cottonseed oil (the lard used in all these specimens had been filtered) and also one specimen of lard very carefully filtered; the cottonseed oil had not been filtered. These several samples were placed in front of the spectroscope so that the light would pass through the liquids in succession. The observation showed that in the case of the pure filtered lard there was a very trivial obscuration of the whole spectrum; in the more carefully filtered lard scarcely any, as the light passed through them; in the sample of pure cottonseed oil the whole of the upper part of the spectrum, from the upper or blue end down to the space between Franenhofer's lines E. and F., was obliterated. A second prism was then adjusted so that the light from another source could pass through the spectroscope, revealing one spectrum above another. These lights were so adjusted as that both had the same degree of brilliancy. In this way they could observe in one spectrum the light passing through unobstructed and in the other passing through the specimen. Examined in this way the filtered pure lard appeared as clear and brilliant in the one spectrum as the unobstructed light did in the other; no difference could be detected. The pure cottonseed oil being brought in the place of the pure lard, only one-half of the spectrum could be seen, the part only through which the light passed unobstructed, the other half being entirely obliterated. The 10 per cent. mixture of cottonseed oil produced a very perceptible obscuration; the 20 per cent. mixture much more, so that an approximate estimate can be made by the degree of obscuration; it would, perhaps, be a rough estimate, but you can certainly say whether it is present or not. The samples of Fowler lard 1, 2, and 3 were, after being heated and filtered, subjected to this test, and examined in the same way; with them there was not the slightest obscuration of the blue end: the one spectrum had the same brilliancy as the other. He has not examined a great variety of oils with the spectroscope, and can not say what, if any, other substance would similarly affect the light. Olive oil affects it differently; that produces a dark band on the lower part of the spectrum. These experiments were specially with reference to the detecting of cottonseed oil, and his experiments have been sufficient to warrant him in claiming that he can certainly detect 10 per cent. of cottonseed oil mixed with lard; he tried it with an admixture of 5 per cent., and found some obscuration; his son claimed he could detect 5 per cent. every time, but he (the witness) will not assert that he can do that, but he can a 10 per cent. mixture; he does not claim to be able to state that the adulteration is certainly cottonseed oil, but where there is no obscuration he will say that cottonseed oil is not present, at least not to the extent of 10 per cent.; with practice, he thinks it probable that a much less adulteration, by cottonseed oil, than 10 per cent. can be detected with certainty. It is necessary that the lard should be melted and filtered, so that all the particles of membrane can be removed from it. If melted and not filtered, there will be a slight obscuration, but it will affect all parts of the spectrum alike; if the lard is properly filtered there will be no obscuration whatever; the presence of cottonseed oil affects the blue end only, and the degree of obscuration, from a slight darkening to a total obliteration, depends on the amount of the cottonseed oil present in the specimen. The micro-spectroscope developed the same effects as the spectroscope alone, in respect to cottonseed oil.

He has tried the elaidine test, but has not been able to successfully use it in detecting cottonseed oil. He has also, repeatedly, tried the Blythe pattern process, but has not, by it, been able to reach any results upon which he could rely.

The testimony for the defense being all in, Dr. Delafontaine was called in rebuttal and made the following statements: *

Prof. M. Delafontaine recalled by the prosecution in rebuttal of statements and theories presented in the evidence of witnesses for the defense, testified that it was evident to him that the scientific witnesses for the defense could not have read his evidence with care, or they would not have charged him with using kettle-rendered lard alone as his standard sample for comparison, as he had distinctly stated that he got a sample of prime steam lard from a packing-house, which, on being tested in the same way as the others, gave about 3 per cent. of residue, and he had added, that, as a clincher, he had taken lard stearine, 1 pound of which is equal to 2 pounds of lard, and found that it did not run higher in residue than his sample, No. 1, of Fowler lard. In other analyses of four samples of prime steam lard of undoubted purity, none ran higher than 3 per cent. of residue. The chemists whose evidence on behalf of the defense, while seeking to impress the board with the unreliability of the process he described in his evidence in chief, all admit they did not, in their attempted trials of that process, follow the process he described, neither faithfully nor closely, as they should have done; all introduced some modifications, some of which are essential, and entirely change the character of the method; others may be of small importance; he can not say whether they are or not. There seems to be a general disposition to raise all sorts of objections to his process; apparently, in the hope that some of them would stick. The gentlemen from the East acknowledge that they have very little knowledge of chemistry of fats; for instance, Professor Doremus says the composition of fats varies a great deal; and reported that he had found 65 per cent. of oleic acid (which means 68 per cent. of oleine) in pure lard; and he (Professor Doremus) says Muter found 47 per cent. of oleine in lard, and claims there is that range of variation in pure lard. Professor Doremus has never analyzed but one sample of pure lard, while he (the witness) has analyzed many, and knows better than Professor Doremus does within what limits pure lard varies. The sample which Muter analyzed was, in all probability, refined lard; he himself, some two years or so ago, analyzed some Chicago refined lard, and found only 48 per cent. oleine; it was mixed with tallow and other things, such as are put into refined lard.

He regards Professor Doremus's reported analysis of four samples of lard—two of which yielded about 95 per cent. of fatty acids, and the other two about 92, or 92.5 per cent. as faulty; because it has been recognized since 1816 that the yield of fatty acids in various fats is nearly the same in all—between 95 and 96 per cent. The man who has done most to enlighten us in the knowledge of fats finds that, "in human fat the total amount of fatty substance in 100 is 95; in mutton fat, 95.5; in beef fat 95; in pork fat 94.9." He (the witness) claims that whenever a chemist finds even 1 per cent. less than these figures he should conclude his analysis is not correct.

He thinks that the explanation of the fact, if it is a fact, that by his process these gentlemen got a less yield of residue, from mixtures of lard and tallow, than from pure tallow alone, is in the time the substances remained mixed before being analyzed; that might make an essential difference. In his tests he took samples of the Fowler lard, and of pure lard, and added to each a quantity of beef stearine; the result, in both cases, was that a residue was obtained equal to what was in the lard, plus what was in the beef stearine. In the case of the elaidine test, all of the gentlemen say it gives no results, but none of them applied it as he testified he did; all had some modification—some used nitric acid.

He has tried the sulphuric acid color tests on the Fowler samples of lard, and got those colors the gentlemen spoke so enthusiastically about, but in treating by that

---

* Op. cit. pp. 267-269.

test he separated the oil, and tested it, which they did not do; if they had done this they would have got the cottonseed oil colors they say was absent in their treatment of it. The cottonseed oil becomes so diluted when mixed with the mass of lard that it is more difficult to detect it in the lard itself than in the separated oil. He did not get the color from the Fowler lard, but he did get it when he treated the oil from that lard alone. He has tried this test recently on the oil from a sample of lard which contained cottonseed oil, and he got the colors they describe and then tried it on some prime lard oil and got no such color. The color test by sulphuric acid, if the oil is treated, is a valuable test for cottonseed oil, but his experiments show that the cottonseed oil may be so well refined that it will not answer to the test quite so well as for oil in the less refined state.

He has tried the process of obtaining crystals for microscopic examination by a solution. He took a sample of the Fowler lard and a sample of pure lard for a comparison, and treated them by that method, and after four or five hours, when half the ether was evaporated, there was in the Fowler lard an abundant deposit of crystals, in the pure lard none yet; he examined those crystals from the Fowler lard, and found them to be nearly all crystals of stearine, with a small sprinkling of crystals of palmitine; the sample of pure lard was allowed to stand four hours longer, the ether then being reduced to one-quarter of its original bulk; from this he got a small crop of crystals; on examining these by the microscope he found them to be mainly crystals of palmitine, with some of stearine; this process leads to determining whether the crystals are those of stearine or of palmitine, when it is carefully applied; the gentlemen on the other side all acknowledge they do not know the difference between the crystals of stearine and those of palmitine; his treatment by this process showed that the Fowler sample of lard had a great deal more stearine in it than the pure lard had.

He thinks the gentlemen have greatly exaggerated objections to the method pursued by Mr. Hoskins in testing by the pattern process; he has seen that process tried on samples of lard and of tallow, and on mixtures of lard and tallow, and found quite different patterns produced in these specimens, when the process was properly applied.

He has tried Husson's method, but did not get satisfactory results from it; he had no difficulty in understanding what was meant by 90 degrees alcohol and 66 degrees ether: both mean the degrees measured by the hydrometer, and are the equivalent of per cent.; it is a quite usual mode of expressing the strength of such substances in French.

DECISION OF THE BOARD OF TRADE.*

The board find that the charges preferred may be properly summarized under the following general heads, to wit:

First.—That a certain lot of 250 tierces of lard, manufactured by the Anglo-American Packing and Provision Company, branded "James Wright & Co. Prime Steam Lard" and marked "⟨69⟩ 10," which lot of lard was stored in a provision warehouse of the Anglo-American Packing and Provision Company, and represented by a warehouse receipt issued by said company—having been by it put upon the market and sold as and for prime steam lard—was delivered to complainants in the course of business and paid for by them as prime steam lard, but was not in fact prime steam lard, as required by the rules of the Board of Trade, but was adulterated and contained substances other than hog lard, to wit, tallow, vegetable oils, etc., as was said to have been stated by competent and skilled chemists who had analyzed the same, and as said complainants charge and believe to be the fact.

Second.—That on the first day of June, 1883, said Fowler Brothers tendered to said complainants a certain lot or lots of lard—brands and marks, other than prime steam lard, not stated—which tender purported to be in fulfillment of a contract made by

---

* Op. cit., pp. 270-272.

said Fowler Brothers to deliver to complainants a large quantity of prime steam lard, then deliverable on said contracts, which said lard, complainants charge, was not in fact prime steam lard, as required by the rules of the Board of Trade, but was mixed and adulterated by and for said Fowler Brothers, with tallow, beef fat, cotton-seed oil, or other substance different from hog's lard; which tender was intended by the said Fowler Brothers to deceive, defraud, and cheat complainants by delivering to them a spurious commodity under the brand and name of prime steam lard.

Third.—Complainants charge upon information and belief that the Anglo-American Packing and Provision Company, with the knowledge and consent of said Fowlers, has manufactured a large quantity of adulterated lard and mixtures which has been sold by said Fowlers to the trade, and to members of the Board of Trade, for prime steam lard, which said adulterated lard is stored in the warehouses under their control, and which transactions complainants charge to be acts of bad faith and dishonorable and dishonest conduct in business.

The board of directors have given to the investigation of these charges a very protracted and patient hearing, which in their judgment has been exhausted in developing all the facts attainable in respect to them, and have arrived at the conclusion that they have not been sustained, and have therefore voted that they be dismissed.

Inasmuch, however, as these charges involve questions of the greatest concern to the members of this association, and to dealers and consumers of pork products, not only throughout our own country but in foreign lands as well, the board of directors, in view of the evidence submitted in this case, both on the part of defendants and for the prosecution, can not, with a due regard to their responsibilities to the public and to the members of this association, refrain from expressing their unqualified disapproval of and censure upon defendants for the remarkable methods of conducting the business of manufacturing lard in their establishment, as developed by the evidence in this case. It appears, and is admitted, to have been the practice, during at least several of recent months, that beef product in various forms has been rendered in the same tanks and with hog product, this mixed product of certain tanks being conducted through a system of intricate machinery and pipes in which also prime steam lard was at times conveyed to their so-called lard refinery wherein both prime steam lard and the mixed product used for what is called refined lard is drawn off into packages for market; and this in a manner that by accident or design on the part of the employés of the establishment could easily contaminate the purity of their prime steam lard, which might thus become more or less adulterated, not only with the beef product so rendered with a portion of their hog product, but also with the cotton-seed oil and other unknown substances used in the manufacture of their so-called refined lard; and this board, in view of the existing methods of manufacturing prime steam lard in this establishment, recommend that, without delay, the parties so readjust their lard-manufacturing arrangements that all grounds for suspicion in this respect shall be effectually removed, and that in case this recommendation is not promptly complied with to the satisfaction of this board, such action be taken as will relieve this board of all responsibility in respect to such product.

The board of directors would embrace this occasion to express their gratification that, as the result of this investigation, the question of ascertaining the truth as to adulterations in lard by scientific examination, which has hitherto, to say the least, been one of extreme difficulty, seems now to give promise of a satisfactory solution; and while not desiring to express absolute confidence in any particular method for determining adulterations by the substances suggested in the charges preferred in this case, the board feels great encouragement to believe that even small adulterations with cotton-seed oil can be detected by some of the methods detailed in the evidence submitted in the case by scientific gentlemen; and that the microscope, in the hands of an experienced operator, can be successfully employed in detecting adulteration by beef product when it exists to the extent of 10, and probably even a much less percentage.

ADULTERATION OF AMERICAN LARD.

IMPORTANT PROSECUTIONS.*

At the Liverpool police court on the 20th ultimo, before Mr. Raffles, several wholesale provision merchants in Liverpool were summoned for having sold lard not of the nature, substance, and quality demanded by the purchaser. Mr. Marks appeared to support the summonses on behalf of the Health Committee of the Corporation. The first case called was one in which Cuffey Brothers, Victoria street, were summoned, and for whom Mr. Pickford appeared. The court was crowded with representatives of the provision trade.

Mr. Marks, in opening the case, said the defendants, who carried on business at 40 Victoria street, were summoned for selling lard which was not of the nature, substance, and quality demanded by the purchaser. The warehouse of the defendants was visited on the 14th May by Inspector Baker, an officer under the sale of food and drugs act. He there saw a number of buckets of lard on which was printed "N. K. Fairbank & Co., refined lard, Chicago." He inquired the price, and ultimately purchased a bucket for 10s. 8d. He gave the usual notice about requiring the purchase for analysis, and offered to divide it for that purpose. His offer was accepted. He then left the defendants a sample, another portion he took to Dr. Campbell Brown, in the usual course, and the third he retained. Dr. Brown furnished a certificate, upon which, as a rule, the case rested. Owing, however, to communications that had been made to him (Mr. Marks) by a gentleman instructed on behalf of the defendants, in this and other cases, it was thought desirable that Dr. Brown should be in attendance, so that he might give evidence in a more ample manner than would appear from his certificate, and in order that the defendants might have an opportunity of cross-examining him. The certificate which Dr. Brown had furnished stated that he had analyzed the lard, and that in his opinion it contained considerably more than 40 per cent. of a mixture of cotton seed oil, and either mutton or beef fat. The court would probably gather this was a case of greater importance than cases under the sale of food and drugs act usually were. Certainly if a court was to be troubled with all the information which had been furnished to him officially, and also anonymously, he should imagine that, so far as the United States was concerned, the people were at present given up entirely to the lard question. [Laughter.] However, in the provision trade the case was undoubtedly regarded as of very great importance, and he was bound to say it was important to more classes of persons than one. It was of vast importance to the makers of lard, whose profits were simply enormous, and also to the consumers of the lard. It seemed that about eighteen months ago it was discovered that lard was being imported into this country which was adulterated. It was imported from America, and the fact appeared to have become known in America, and to have created a tremendous amount of feeling there.

Mr. Pickford objected to these observations as being irrelevant.

Mr. Marks said he was simply leading up to the facts. What he was going to say was, that the principal manufacturers of this lard were persons whose names appeared upon the buckets, namely, Fairbank & Co., of Chicago, and Armour & Co. It was known to Dr. Campbell Brown, about eighteen months ago, that this importation of lard was going on.

Mr. Pickford again objected to Mr. Marks entering into matters not connected with the present case.

Mr. Marks contended that his observations had reference to the case before the court.

Mr. Raffles said he should rule that what Dr. Brown did eighteen months ago was not relevant, except it had reference to the summons now being heard.

---

*The Analyst, July, 1888, pp. 136 et seq.

Mr. Marks (continuing) said that in order to meet Mr. Pickford's objections, he would put the matter in this way: On examining the sample in question, Dr. Brown found that his researches, which had occupied him more than eighteen months ago had furnished him with knowledge that had enabled him to discover that the sample contained cottonseed oil. All he was going to say before was that eighteen months ago Dr. Brown could not have done so. [Laughter.]

During the last month, when Dr. Brown analyzed the sample, he found there was a considerable proportion of cottonseed oil, and also of stearine, either mutton or beef fat. Lard adulterated in that way, Dr. Brown would show, was a very inferior article indeed, was not of the same value, and was not as useful for the purposes for which genuine lard is used. The cottonseed oil produced in one season in the United States amounted, he believed, to 150 or 200 million pounds weight. Of that a very large percentage, pretty nearly half, was used by manufacturers of what was called refined lard, but which the prosecution suggested was adulterated lard. The price of cottonseed oil was only about 22s. 6d. a hundredweight, and the price of beef fat only 30s. a hundredweight, whereas the price of pure lard was 42s. 6d. a hundredweight. Therefore it would be seen that when such large quantities of cottonseed oil were used, enormous profits resulted to the manufacturer, who could substitute for the more valuable article a cheaper one. Every week there came into Liverpool £20,000 to £30,000 worth of lard, and the whole of that he presumed was used in the preparation of food. It was consequently a serious matter for the consumer, as well as a matter of considerable importance to the fair traders in lard, who offered for sale the genuine article.

Inspector Baker deposed to visiting the defendants' warehouse, and purchasing a bucket of lard on which was the name of Fairbank & Co.

Mr. Pickford did not cross-examine.

Dr. Campbell Brown, examined by Mr. Marks, stated that on the 15th of May he received the bucket of lard from the last witness and analyzed it. The result of his analysis was that he found the lard to contain a very large quantity of cottonseed oil and beef or mutton fat. He estimated the total quantity as considerably more than 40 per cent. He really believed the quantity was more than 50 per cent., but he was certain that it was more than 40 per cent.

Cross-examined by Mr. PICKFORD:

Q. I suppose you mean the cottonseed oil and the beef or mutton fat together made 40 per cent.?—A. Yes.

Q. You don't distinguish the one from the other?—A. I have not done so.

Q. Can you do so?—A. I can't tell the precise quantity of beef fat.

Q. Then, I assume, if you can't determine the quantity of beef fat, that you can't determine the quantity of cottonseed oil?—A. I am quite certain there was more than 30 per cent. of cottonseed oil, but I can't determine more. I can't estimate exactly the quantity of beef fat, and therefore I can't tell precisely the quantity of cottonseed oil over 30 per cent.

Q. Does not your test tell you anything about the beef-fat stearine?—A. It tells a good deal about it, but not the precise quantity.

Q. How do your tests show the presence of these things?—A. I think you would require to attend a course of lectures on chemistry before you do it. [Laughter.]

Mr. RAFFLES. I think we had better not have that.

By Mr. PICKFORD:

Q. What kind of tests do you use?—A. I put the whole thing through seven or eight processes and then argue the thing out. I have no individual test.

Q. Supposing there was no cottonseed oil at all?—A. Yes, but there is. [Laughter.]

Q. Supposing there was no cottonseed oil, but beef fat only added to the lard, could you distinguish that fat from the other fat?—A. I can distinguish beef fat from hog's fat.

Q. These are, as a matter of fact, new tests?—A. They are new applications of old knowledge.

Q. I mean that you could not until quite recently distinguish cottonseed oil or beef fat in the lard?—A. I was quite certain about the cottonseed oil fifteen months ago, but I did not see my way to get out the quantity sufficiently for judicial purposes. Four or five years ago I knew about a certain quantity of beef fat. I don't want to give my results now, for two reasons. One is that I am getting the quantity less every week, and the other is I don't want to let the makers of lard know how little they can put in without detection. [Laughter.]

Mr. Pickford, for the defense, said his worship would probably have divined from the cross-examination that he was not going to deny the presence of cottonseed oil in this refined lard, but he was going to say that "refined lard" was a perfectly well-known trade term, and everybody was aware that it was a compound of fats and not real hog fat. If that was proved, and the effect of the notice on the barrel that it was refined lard meant what he said, then he should bring the case within the case which was decided without mustard, in which a man asked for mustard, and on the packet was a notice that the contents were not pure mustard, but compound mustard. He did not think he was wrong in trying to stop what seemed to him to be irrelevant statements—of which a number had been made—about the comparative prices of the ingredients.

Mr. MARKS. I can prove my statement if my friend wants.

Mr. PICKFORD. A great number of these statements are absolutely inaccurate.

Mr. MARKS. I think it is only fair to myself to say what I base my statements upon.

Mr. RAFFLES. I have nothing to do with it.

Mr. MARKS. I was basing my statement upon the general broker's trade circular of the 15th of last month.

Mr. RAFFLES. Don't let us talk about it.

Mr. PICKFORD. No, sir; I don't want to. But if I allow it to pass without contradiction it would no doubt be stated in the papers that it was admitted as a fact, and I don't admit it. I do admit that this refined lard is sold at about the same price as pure lard, and I admit that as being important for showing that the people who use a very large quantity of this stuff in this country willingly pay the same price, and have no complaint to make of the stuff.

Mr. Pickford was proceeding to make remarks upon the summons when—

Mr. RAFFLES said: The only question I have to deal with is the adulteration.

Mr. PICKFORD. But is it adulterated?

Mr. RAFFLES. On Dr. Campbell Brown's evidence it is adulteration.

Mr. PICKFORD. I say it is not adulteration. It is sold as refined lard, and I can call evidence——

Mr. RAFFLES. I can't go into that.

Mr. PICKFORD. If you say you can't hear evidence on that I can't say any more.

Mr. RAFFLES. I can't.

Mr. PICKFORD. But supposing that the meaning of the trade term "refined lard" is not pure hog's lard, then it is a compound and not adulterated.

Mr. RAFFLES. I can't go into any special meaning which is attached by the trade to refined lard.

Mr. PICKFORD. If the words "refined lard" do mean a compound of fats, it means that everybody who buys the stuff, that the purchaser is not getting an article of a different nature, quality, and substance to that demanded——

Mr. RAFFLES. Well, of course, this case will go elsewhere, whatever my decision is.

Mr. PICKFORD. Then I understand that you reject any evidence on my point?

Mr. RAFFLES. Yes.

Mr. Pickford said he ought perhaps to state that he intended to show that refined lard was a well-known term, meaning a compound of fats and oil, and was not confined only to the American market, but the English refiners, he believed, used some cottonseed oil. Refined lard, whether American or English, was never pure hog's fat, and everybody knew it.

Mr. RAFFLES. Dr. Campbell Brown tells me this is adulterated, and that is quite sufficient for me. He will take the case elsewhere, I suppose.

Mr. PICKFORD. It is very likely, but I can't say anything one way or the other.

Mr. RAFFLES. It is a very serious question, and one which ought to be dealt with seriously. I shall inflict a fine of £5 and costs.

Messrs. Pelling, Stanley & Co., Victoria street, were summoned, and Mr. Mulholland appeared for the defense.

Mr. RAFFLES. What is the difference between this case and the last?

Mr. MARKS. An important one, in one respect, inasmuch as the lard is the same manufacture, but is from the other great refiner's. This also is called refined lard, but whether the defense means to rely upon that or not I don't know.

Inspector Baker gave evidence of the purchase of the lard from the defendants, and stated that when he offered to leave a sample with them they declined to receive it. He then took the lard to Dr. Campbell Brown.

Dr. Campbell Brown said he analyzed the lard in question and found a very large quantity of cottonseed oil and fat extracted from beef or mutton. He estimated the total quantity, approximately, at 40 per cent. The case was not so bad as the last.

Mr. Mulholland said that the witness had analyzed two samples of lard of the same brand, and had given different figures. He therefore applied for the case to be adjourned for the lard to be sent to Somerset House for analysis.

Mr. Raffles said he should decide the case himself.

In cross-examination Dr. Campbell Brown said he believed that of late years the oil formerly taken out of hog's fat to make it harder and more fit for carriage was not now salable because of the introduction of mineral oils for lubrication.

Mr. MULHOLLAND. And therefore it becomes commercially more important to put hardening stuff into the lard than to take out the oil?

The WITNESS. It becomes important to the manufacturers, but it is not right.

Mr. MULHOLLAND. You are not the analyst of right and wrong, but only of fat and stearine. [Laughter.]

Mr. Mulholland submitted that the case came within the first subsection of the act, namely, that these substances were not introduced for any fraudulent purpose, but for the purpose of hardening the lard and making it merchantable.

Mr. RAFFLES. It can not be for the purpose of bringing it over. I am against you there.

Mr. Mulholland said he also took the same point as Mr. Pickford, and as his worship was against them there he supposed he would grant them a case on that as well as on other points if necessary.

Mr. RAFFLES. I will give you any means you may require for testing my decision. It is very desirable this matter should be settled. I shall inflict the same penalty of £5 and costs.

Messrs. J. & T. Edwards, provision dealers, Whitechapel, were summoned for a similar offense.

Mr. Edwards, a member of the firm, appeared and stated that he bought the lard as pure lard.

Mr. MARKS. If you did I should sue the manufacturers for the amount of the fine and costs.

The invoice for the lard was handed to his worship, who, however, said it was no warranty of purity.

Mr. Edwards said he bought the lard as pure, and consequently he thought the fine ought to be less than in the other cases.

Mr. Raffles said he should inflict the same fine, £5 and costs.

SUMMARY.

I have endeavored to set forth in the preceding pages our present knowledge concerning the constitution of pure lard and its adulterations. The question of the wholesomeness or unwholesomeness of the various ingredients has not been raised in these investigations. It is hardly necessary to call attention, however, to the fact that the stearines and cotton oils used in the manufacture of adulterated lard are, so far as known, perfectly wholesome and innocuous. There is every reason to believe these are fully as free from deleterious effects upon the system as hog grease itself.

A more serious question which is presented is the effect of selling adulterated lard as pure lard or refined lard. To do this is a fraud upon the consumer. Although it has been claimed by the large manufacturers of refined lard that the term refined is a trade-mark whose meaning is perfectly well known by seller and purchaser, yet it can not be denied that the meaning of the word refined in the above sense is generally unknown to the consumer. The idea conveyed to the ordinary consumer by the word refined would be an article of superior purity for which he would possibly be willing to pay an increased price. It is gratifying to know that since the investigations recorded above were commenced the largest manufacturers of compound lard in this country have decided to abandon the use of the term refined and to sell their lards as compound lard or lard compounds, and, in cases where no hog grease at all enters the composition of the article, to place it upon the market as cottolene or cotton-seed oil product.

In the cases before English courts cited above it is seen that the word refined does not convey to the judicial mind the idea which is claimed for it as a trade-mark, and hence the wisdom of the manufacturers in changing the labeling of their wares is at once manifest.

The extensive adulteration of American lards has afforded grounds to foreign countries for prohibiting importation of our production or of levying upon it a heavy duty. By requiring all food products made in this country to be labelled and sold under their true name we could secure for our products immunity from any such exclusion from foreign countries as is mentioned above. The right of foreign countries to levy an import duty on our products is one which we would in no measure seek to abridge; yet by the recognized purity of our exported food articles we should see that they secure a proper entrance into foreign countries. These remarks are not alone applicable to lard and its adulterations, but to all kinds of food products, whether they are to be consumed at home or abroad.

INDEX.

A.

	Page.
Acids from cotton oil, composition of	508
Adulterants in lard, calculation of, by Brullé's test	476
from iodine absorption	472
melting point	475
refractive index	475
formula for	474
rise of temperature with sulphuric acid	475
formula for	475
specific gravity	470
formula for	471
determination of, by chloride of sulphur	475
quantitative determination of	469, 470, 471, 472, 473, 474
Affidavits, samples of	477, 478, 479
Allen, A. H., cocoa-nut oil in lard	506
detection of cotton oil by	507
recommendation of	465
Analysis, methods of, employed	428
Analyses, results of	477
Armour & Co., percentage of adulteration in lards of	476

B.

Bechi's method	465
Beef fat	410
Belfield, Dr. W. T., certificate of	525
Bizio, criticism of Bechi's test by	501
Board of Trade, decision of, in the case of McGooch, Everingham & Co., against Fowler Bros	541, 542
Brown, Prof. J. Campbell, errors of analysts	511
Brullé, method of	502

C.

Carr, Mr. Oma, method of	448
Chicago Board of Trade, definition of	406
Cholesterine and phytosterine, occurrence of, in glycerides	514
Chloride of sulphur test	468
Cocoa-nut oil, occurrence of, in lard	506
Color reaction	415
method of determining	448
Computing specific gravity	434, 435, 436, 437
Conroy, Mr. Michael, tests for adulteration of lard	513
Cotton oil	411

	Page
Cotton oil, chemical properties of	421
color reaction of	420
crystallization point of fatty acids in	420
iodine number of	421
manufacture of	413
melting point of	420
fatty acids in	420
mills, location of	411
physical properties of	418
other properties of	421
reaction of, with nitrate of silver	421
refractive index of	420
rise of temperature of, with sulphuric acid	420
saponification equivalent of	421
Specific gravity of	418
at different temperatures	419
volatile acids in	421
weight of, used in mixed lard	428
Crampton, Dr. C. A., work of	434, 469
Crystallization point of fatty acids, method of determining	446, 447

D.

Danforth, Dr. I. N., certificate of	526
Delafontaine, Prof. M., rebuttal testimony of	540, 541
testimony of	517, 518, 519
Doremus, Prof. R. Ogden, testimony of	537, 538, 539

E.

Elaidine reaction	515

F.

Fairbank & Co., percentage of adulteration in lards of	476
Formulæ for lard analysis	527, 528

G.

Grease, pig's-foot	409
white	409
yellow	409
Grey, Mr. Watson, iodine absorption	511, 512
Gibson, Prof. C. B., testimony of	522
Guts, definition of	406

H.

Habershaw, Mr. W. M., certificate of	536
Haines, Prof. Walter S., certificate and methods of	529, 530
Hayes, Prof. Plymmon S., certificate of	526
Hehner, O. H., opinion of Bechi's test	509
Hesse, discovery of phytosterine	514
Hirsch, Mr. J. M., testimony of	521
Hirschsohn, experiments with chloride of gold	502
Hog-fat products	408
Horn, Mr. M. F., estimation of oils, etc., in mineral fats	512
Hoskins, Mr. William, evidence of	519, 520

INDEX. 551

I.

	Page.
Iodide of potassium solution	464
Iodine number, method of determining	462
obtaining	464
re-agents for	462
solution, variation in strength of	463
Isbert & Venator, separation of stearine and palmitin	515

J.

Jones, studies in lard, adulteration of	510

L.

Lard, adulterants of	423
properties of	413
and compound lard, comparison of properties of	427
analysis, abstract of methods of	516
butchers'	407
chemical properties of	417
choice	405
kettle-rendered	405
crystallizing point of fatty acids in	416
definition of	405
exports of	428
fixed acids of	417
free acids of	417
industry, statistics of	427
iodine number of	417
leaf, definition of	405
melting point of	414
fatty acids in	417
microscopical appearances of	418
mixing	414
moisture in	418
natural, definition of	405
prime steam	406
pure, properties of	411
reaction of, with nitrate of silver	418
refractive index of	416
rise of temperature with sulphuric acid	416
saponification equivalent of	417
specific gravity of	414
volatile acids of	417
adulterated, chemical properties of	426
color reaction of	424
crystallization point of fatty acids of	425
iodine number of	426
melting point of	424
fatty acids of	425
microscopical appearances of	426
moisture in	426
physical properties of	424
reaction of, with nitrate of silver	426
refractive index of	425

	Page
Lard, adulterated, rise of temperature with sulphuric acid	425
saponification equivalent of	426
specific gravity of	424
volatile acids in	426
Lards and oils, classification of	477
Letter of submittal	403

M.

Mason, report of	502, 503, 504, 505
Melting point, determination of	439, 440
of fatty acids, method of determining	448
Micrographic plates, description of	451, 452, 453, 454
Microscopic examination, method of procedure	449, 450, 451
Milliau's method	466, 467
Moerk, lard test of	502
Munroe, Prof. C. E., letter from	445
experiments of	446
Mutton tallow	410

O.

Oil, white	412
Oils and lards, classification of	477
Oleo oil	410
Oxygen, absorption of, by cotton oil	515

P.

Palmitine and stearine, separation of	515
Picnometer	428, 429
Phytosterine and cholesterine, occurrence of, in glycerides	514
Prosecutions for adulteration of American lard	543, 544, 545, 546
Pure lards, table of composition of	480

Q.

Qualitative reactions	516

R.

Reaction with nitrate of silver, method of determining	465
Reactions, some peculiar	465
Re-agents, preservation of	462
Reducing power of cotton oil, loss of	467
Refining process	412
Refractive index at temperatures above normal	442, 443
method of determining	441, 442
Remsen, Prof. Ira, certificate and method of	532, 533, 534, 535
Rendering-tanks, description of	408
Rise of temperature, method of determining	443, 444, 445
Rose, Dr. P. B., method of	528
testimony of	516

S.

Salkowski, test for cholesterine and phytosterine	514
Saponification equivalent, method of determining	461, 462
apparatus for conducting	459

	Page.
Sausage casings	407
Sharpless, Prof. S. P., certificate and explanation of	531, 532
test for adulteration of lard	512
Soda, thiosulphate of	463
Soluble and insoluble acids, determination of	455, 456, 457, 458
fatty acids, calculation of	458
Specific gravity, method of	428
of fats in a solid condition	433, 434
Spectroscopic examination of fats	516
Sprengel's tube	434
Starch paste	464
Stearic and oleic acids, determination of mixture of	449
Stearine, lard	409
oleo	410
and palmitine, separation of	515
Stearines	409
chemical properties of	422
color reaction of	422
iodine, number of	423
melting-point of	422
microscopical appearances of	423
moisture in	423
physical properties of	422
reaction of, with nitrate of silver	423
refractive index of	422
rise of temperature of, with sulphuric acid	422
saponification equivalent of	423
specific gravity of	422
use in lard adulteration	421
volatile acids in	422
Stewart, certificate and method of	531
Stock, modification of Milliau's method by	510, 511
Summary	547

T.

Table No. 17	480
description of samples in	481
notes on	481
No. 18	483
description of samples in	482
notes on	484
No. 19	485
description of samples in	484
notes on	486
No. 20	487
description of samples in	486
notes on	488
No. 21	489
description of samples in	488
notes on	490
No. 22	492
description of samples in	490, 491
notes on	493

	Page.
Table No. 23	495, 496
description of samples in	493, 494
notes on	497
No. 24	498
description of samples in	497
No. 25	500
description of samples in	499
Tallow, composition of	508
Tilley, Dr. Robert, testimony of	523, 524

V.

Volatile acids, determination of	455, 456

W.

Wallace, Dr. Shippen, experiments of	501
Warren's test	468
trial of	469
Weighing fats, general directions for	460, 461
Wesson, David, experiments of, with Brullé's test	506
with lard tests	502
microscopic tests for adulteration of lard	513
Westphal balance	430, 431, 432
Wheeler, Prof. C. Gilbert, testimony of	523
Williams, Mr. Roland, iodine number of fatty acids	512
studies in lard adulteration by	509, 510
Witthaus, Prof. R. A., certificate and method of	535, 536